THE KLINK

GOOD DAYS AND BAD
TRIBUTARIES OF THE
JUNE DAY
SUMMER
ON THE LOWER

First published in Great Britain by Merlin Unwin Books Ltd 2024

Merlin Unwin Books Ltd
Palmers House
7 Corve Street
Ludlow
Shropshire SY8 1DB
UK
www.merlinunwin.co.uk

ISBN 978 1 913159 75 7
Edited, designed and typeset in 11.5pt Minion Pro by Merlin Unwin Books
Printed by Bell & Bain Ltd, UK

THE KLINK

The Extraordinary Flytying Geni

HANS VAN KLINKEN

MERLIN UNWIN BOOKS

Dedication: to Ina

Being married for more than 40 years to the same person is not the norm anymore. It has become more of an exception.

I want to take the opportunity now to tell you how proud I am of Ina. She is by far the best example of unconditional love and compassion that I have ever known. I am blessed to be married to a woman who fully shares my love of the outdoors, flyfishing, wildlife, and nature.

I could never have foreseen early on in our relationship that Ina would become not only my best friend in life, but my best fishing buddy too! No fishing trip without her can compare to a trip we do together, and that's without looking at the catches. Many pictures that Ina took of me fishing you will find throughout this book. Several of her pictures have been published as covers of fishing magazines and dozens have been used by tourist departments in Scandinavia and Canada. Most famous by far are many pictures that she took during my kid classes in Canada. Beside her photographic skills, she surely is an expert on fishing the dry fly for Atlantic salmon. Ninety per cent of her fishing for Atlantic salmon is done with dry flies.

Her catches of bonefish and tarpon in the Caribbean are also spectacular. While many of her bonefish reached the 10lb range, I am incredibly proud of the 110lb tarpon she caught and landed in Belize a few years ago.

I am so grateful for her ability to share my interests and provide the additional support required for me to spend all those countless hours in front of the computer which were necessary for me to complete this book and write the many stories and book contributions that have been published. My friends and other fishermen love her for her excellent company during our travels.

Thank you so much Ina; without your love and support, this book would be just another of my 'things' yet to do.

Contents

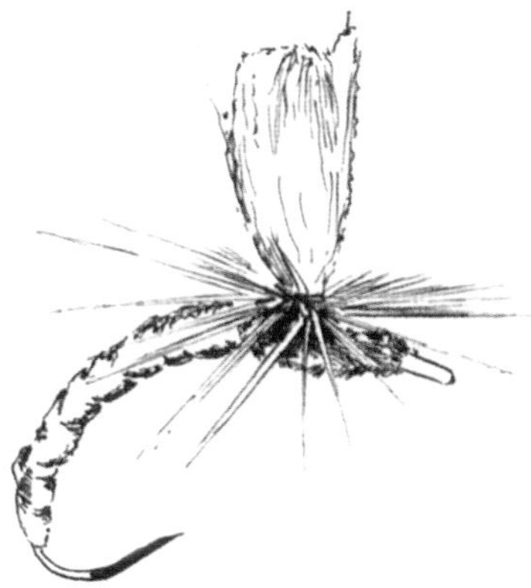

Acknowledgements

To list everyone who helped me with this book would surely fill a second book. These acknowledgements have been one of the hardest parts to write, due to fear of accidentally leaving someone out.

There are many individuals to whom I am especially indebted because without them there simply would not have been any book at all.

First, I would like to thank Dr Tom Sutcliffe. Tom is a South African flyfishing legend, writer and artist. Tom helped me a lot whenever I lost my writing inspiration or was struggling with the English language. Without his help, I would surely be writing still.

Grateful thanks go to Leon Links, John Roberts, Oliver Edwards, James L. Hardy, Arild Blomfeldt, Wayne Luallen, Hein van Aar, Philip White, Kenneth Boström, Gunnar Johnson, Tomas Olsson, Bo Gullefors, Simo Lumme, Juha Pusa, Lauri Sälevä, Antti Sorro, Matti Rantanen, Antti Vanninen, Juhani Kyllönen, Anne Hyytiäinen, Simo Yli-Lonttinen, Anni Yli-Lonttinen, Akki Rinne, and Ari Savikko, who all have helped me enormously with the inspiration, motivation, ideas, translations and extra information to make it possible to make this a very special book indeed.

My gratitude also goes to Hans de Groot, Franklin Moquette and Ton Lindhout who came up with the amazing Klinkhåmer name for my best fishing fly.

American flyfishermen and flytyers like Lee Wulff, Eric Leiser, Dick Nelson, Al Worthington, Leon Chandler, Poul Jorgenson, John Betts, Gary LaFontaine, Dick Talleur, Ron Alcott, Dick Lemmerman, Al Caucci, John Randolph, Marvin Nolte, Maxwell MacPherson, Mike Martinek, Chuck Furimsky, Vince Marinaro, Doug Swisher, Carl Richards, Gary Borger, Darrel Martin and Dave Whitlock all had an enormous influence on my flytying development over the past 40 years.

Thank you Louis Noble, Tony Deacon, Jim Gourlay, John Davison, Lizelle Jacobs, Tom Sutcliffe, Glenn C. Smith and Steven Brown who supported my attempts in writing and were an enormous help in improving my English until I finally had enough confidence to write this book.

Thanks too to all members of the Grayling Society in the UK, Holland and abroad with whom I share my intense passion for grayling.

Huge thanks to Ina van Klinken-Stevens, Leon Links, Hein van Aar, Antti Vanninen, Aki Rinne, Juha Pusa, Tomas Olsson, Arild Blomfeldt, Lauri Sälevä, Arikko Miettinen, Barry Ord Clarke, Branko Gašparin, Roy Christy and Antti Sorro for generously providing me with their wonderful pictures, awesome flies or special artwork and permission for their use.

Thank you, Karen, Lydia, Jo and Joanne from Merlin Unwin Books, for believing in my project, for your great advice, outstanding help and our marvellous collaboration throughout this book.

I would like to offer special thanks to Peter Rosmalen, Alan Bramley, Terry Ruane, Ole A. Bjerke and Bill Chase for the excellent cooperation and production of my special hooks needed to create my Klinkhåmer Special in the past, present and future.

A special word of thanks to Bill Stephens, Lee Wulff, Barbara Genge, Junior Flynn, Gordy MacKinnon, Paul Gale and Ruth Forsey-Gale who are directly responsible for widening our skills and experience with dry fly fishing for Atlantic salmon in Eastern Canada.

Last but not least, my gratitude goes to Barbara Genge (Tuckamore Lodge), Ruth Forsey-Gale (Pirates Haven), Laura Farley (Danville), Judith G. McCann (Fly-Rite), the Ruciński family (Renomed) and the Kitchener family (Semperfli) who have sponsored my kid classes over many years.

And a big thank you to all the unnamed people who have played their part in my flyfishing development and flytying passion over the years.

True Friendship

Leon Links and I have been close friends since 1988, which is a very long time indeed. Our close friendship is based on mutual affection, trust, support, and respect for one another. We have no secrets between us and our friendship is based on honesty, integrity, and frank communication. We have always been there for each other through good times and bad, offering a listening ear and a helping hand.

In the late 1980s, we met for the first time and I re-introduced Leon to flyfishing in Scandinavia, a region he had not fished for about twenty years. In the same way it had happened with me, the beauty of the natural surroundings quickly cast its spell on Leon and his passion for flyfishing took hold of him. He started to collect flyfishing books, wrote many articles, built up an enormous circle of fishing friends, and started to design his own flies, many of which were tied with CDC. Leon's skill in taking flytying pictures is outstanding, and many of his macro photographs of flies are among the world's most famous fly images.

Like me, Leon has been a long-time member of the editorial staff of the Dutch flyfishing magazine *De Nederlandse Vliegvisser*, and he has written for magazines in Holland, England, Belgium, France, Sweden, Germany, Finland, South Africa and Italy. In September 2002, he wrote the excellent book: *Tying Flies with CDC*, published by Merlin Unwin Books in England and Stackpole Books in USA. Later it was also published with translations in French and Czech language.

Nowadays, Leon's favourite waters for trout and grayling fishing are those in Scandinavia, Germany and Slovenia. Our friendship is unique. We fished a lot together, tied many flies together, did countless workshops and classes together and wrote dozens of articles together. Most tying steps and fancy fly pictures in this book are taken by Leon in his own macro studio. Leon was a major motivator for me to write and finally finish this book, despite some long delays due to medical and family situations. It was often Leon who helped me to find new inspiration and strength to continue and finally finish this book.

Thanks a million, my friend!

Foreword
The man who changed the way we fish

Does a book by Hans van Klinken need a foreword? I doubt it. His fly, *the* fly, has been on our lips for decades. He is famous throughout the flyfishing world. His Klinkhåmer floats on every trout or grayling river across the world. The fish I've caught on it are numbered in their thousands. You too? Very probably. When Hans asked me to write a few words I happily agreed. After all, I have picked his brains and have stolen his flies for my last half dozen books. It is hardly payback time. What can be offered in return for the gift of the Klink to the flyfishing world?

Many years ago I discussed the subject of Hans and his flies with my friend the late Oliver Edwards after I had made a couple of fishing trips with Hans in Germany and Yorkshire. I had done rather better than I ought and my face-saving was down to Hans' flies. Oliver exclaimed, 'Lad's a bloody genius. That fly is flippin' fantastic. Hans thinks it's a sedge. I've no idea how he did it.' This is not verbatim; the colourful adjectives have been toned down, but it's how I recall Oliver's reaction. Genius. One pithy Yorkshireman had summed him up and I agreed.

Forty years on, thousands of flyfishers across the world have had cause to reiterate those words or at least give heart-felt thanks to Hans for a handful of wonderful trout and grayling flies. And the best of these is the incomparable Klinkhåmer Special.

The Klinkhåmer really did change the way we fished on rivers. Firstly for grayling, and then for trout, and now it seems, for other species too. In the years before HvK, grayling flies were invariably small: the largest in Europe was a size 12 adult caddis. Hans offered a radical alternative, a large dry fly but with two-thirds hung below the surface.

Today it's accepted worldwide but believe me forty years ago at first sight of an inch-long dry fly, anglers would frown, raise an eyebrow, laugh or at least be highly and verbally sceptical. Until they tried it for themselves. It became a fly to give every flyfisher a new tactic, especially in faster water. River flyfishing changed irrevocably.

A mere handful of inventors of flies have had flyfishing immortality bestowed upon them. Lee Wulff and Frank Sawyer both achieved such status with the series of successful flies that bear their names. Every inquisitive flytyer has had a go at improving a tying style or introducing new materials, hoping to be a catalyst in tying evolution. We've all had a shot at fame; some achieving it for this month's edition of a fishing magazine, only to be granted the fifteen minutes Andy Warhol predicted for us. Many fly innovators like Richards, Swisher, Goddard, Shipman, LaFontaine, Fratnik and a few dozen others rightfully have had accolades, often for a pattern or tying style to match a single hatch or a set of circumstances.

But perhaps Hans van Klinken alone has devised in the Klinkhåmer a pattern type that can be colour- and size-matched to represent the widest possible range of trout and grayling food; not that it stopped at just those two species.

For too many years flyfishers across the world have been waiting for Hans to write a book explaining his dry fly philosophy. His dressings are well known but here at last we have a real insight into the how and why of Hans' thinking about the dry fly.

The Klinkhåmer is forty years old and with a few tweaks along the way it remains ninety percent the way it was first designed. The fly I use today is identical to the original tying Hans gave me 36 years ago. But this book is much more than a history of a seminal pattern. Rather, it is about how trout and grayling see surface food, what Hans' responses are to the problems of presentation and imitation, how a simple tying of the appropriate materials can be so effective when fished in the correct manner at the right time. For much of his dry fly fishing, Hans' preference, like my own, is to use a fly with a parachute hackle. Within these chapters there is no better summation as to why the style is so effective.

Like many of his patterns, Hans is larger than life. He is a big man, gregarious and generous, with a unique story to tell. His tying demonstrations and talks have entertained and inspired thousands of anglers across the world. He is never happier than when enthusing young people to flyfish and to tie flies. He owes the fishing world nothing; it is we who are indebted to him.

John Roberts
York 2024

patagonia

First cast
Hans van Klinken

Not so long ago, someone unfamiliar with the sport of flyfishing asked me to try to explain it in just one sentence. Forty years ago, I would have needed three hours or more to explain it, but after nearly five decades of asking myself the same question, I no longer find the answer difficult. In fact, I find the answer to be quite simple. So simple that I only need a few words: ***Flyfishing is merging with nature!***

For me, flyfishing is all about the pleasure and satisfaction that a person can acquire while communing with the surrounding nature and wildlife. When I look at my closest circle of fishing friends, I can say for sure that it never matters if we catch fish or not. Our pleasure comes from just being out there on the riverbank and enjoying the river and its surroundings in all its beauty. From the earliest days of my career as a flyfisher until now, I have progressed through many stages and each individual phase was extremely important in shaping the approach I have to fishing today.

Maybe flyfishing is an addiction or obsession for many, but I don't think it is for me, because I also grab my camera, tying vice or pencil as willingly as my fly rod. I like to chat with fellow fishermen and guides and I can watch beavers, otters, moose, birds of prey and caribou for hours with my fly rod just standing next to me. For me, it's all part of flyfishing. When I see kids struggling with their fly rod I am the first who tries to help them, by tying on a new leader and one of my secret flies in the hope they will catch a fish.

With good friends we can sit on the grass of the riverbanks and talk about flies and forget all about time. Missing a good hatch is not unknown because all these aspects make up such a big part of flyfishing for me!

Gazing over the river, I enjoy the movements of the current, listening to how the wind plays in the treetops and the birds sing in the background. I can sit for hours like this, but admit it has taken me years to reach this level of understanding and satisfaction. Maybe it's a kind spiritual feeling as well. Even though I was close to nature as a young boy, my feelings are a lot different today. Sometimes I am full of sadness about what we humans have already destroyed in just one millennium.

From my earliest days as a flyfisher until now, I have progressed through many stages, and each individual stage was extremely important to the shaping of my flyfishing attitude today. I too had my very long days of fishing and my 'eager to catch as many fish as possible' days in my early years, and that is one reason why I can understand young, fanatic anglers so well. It's natural to want to prove yourself and to want to be sure that your flies will work and there is no better proof than numerous catches. I started as a dedicated dry fly purist for the first ten years of my flyfishing, but with time and experience, I developed my own techniques and flies to be able to catch fish at depths of 4-5 metres.

I am a very passionate river flyfisher but that doesn't mean that I don't like to fish in lakes, ponds, reservoirs or saltwaters. I love all kinds of flyfishing as long it's not too crowded. I can't deny that I feel closest with nature when standing in a remote river anywhere on the globe. I have been alone in the wilderness for long periods and it was important for me to develop myself and my own approach to fishing. As a student with

plenty of spare time and my father's reward for helping him at work, I managed to spend most of my summer holidays in northern parts of Norway, Sweden and Finland, which marked the beginnings of my love affair with the great outdoors from 1971. In 1975 I stayed for almost four months in the wilderness and in this book I explain more about this defining trip. I made a lot of very remote fishing trips in the 1970s and 1980s and all I can say is you love it or you hate it.

The long months experience that I had gained up in Scandinavia and Finland was a great help in my research, exploring waters all by myself. I fished frequently and the more time I spent in these regions flyfishing, the more I learnt.

People sometimes claim that I am a real expert in fishing the mighty rivers in Scandinavia but I am NOT and I am still learning today. Most of the waters I fished in those days were 600-1,000 metres above sea level. The weather conditions were tough with unpredictable mountain winds and an extremely high population of wild, hungry mosquitoes and black flies. I taught myself how to deal with them, but I guess many people would not bother and this will stop them doing some of the best flyfishing in the world. This type of wilderness is what I really love and it is these areas that I believe grayling and trout love the most too! It is a very short fishing season in the Arctic and even sub-Arctic, but a very productive one. The lakes give the fish a very good chance of surviving during the harsh winters and they have an excellent food supply. The insect life is abundant with superb caddis hatches and many species of upwing flies to study. As a result of these experiences I ended up writing about my travels and the flies I developed. It was in these isolated places that I developed the 44 flies of the L.T. series in the 1980s. Several of them later became rather well-known; one of them is of course the Klinkhåmer Special, and number seven from the list has its 40th birthday in June 2024 and that is a big reason for writing this book, to share with you the thoughts behind my flies and how I developed several of the most important ones.

I also love to explore and catch fish species that I have never caught before and that made me enormously flexible, inventive and creative. Fishing so many different areas and waters also widened my flytying experiences enormously and without my travels I never would have tied so many different patterns.

Flytying is an awesome hobby but a lot depends on the flytyer's ability to be inventive and imaginative. A large number of people only tie well-proven flies; others just tie what they need for a successful day's fishing; while another group goes into the serious art of creating the most beautiful realistic imitations imaginable. Flytying has many facets to it, and some people just like collecting tools, vices, feathers, synthetics, silks or furs. For me it's all part of our great hobby and we should respect each other's skills and try to help one another as best we can. Since my retirement from my 'day job' in 2012 I have conducted over 200 workshops or classes with kids worldwide and it only makes my hobby more amazing every day.

The flytying world looks very different today to when I started out. With so many good online tutorials, excellent Facebook flytyers, flytying movies, step-by-step instructions, free e-books and the available materials and quality tools, it has become a lot easier than back in the 1970s and early 1980s. I even know quite a number of flytyers who actually never go fishing.

Flytyers can easily change their specialities or preferences today. One moment they are in the spell of making streamers and the next moment they suddenly love tying emergers, North Country spiders, terrestrials or realistic nymph patterns. Although I went through many stages and even seriously tried classic salmon flytying for seven years, I now concentrate my efforts on tying and developing *effective* fishing patterns, irrespective of the species. I believe that many flytyers have similar journeys.

Flytying means a lot to me. It is also an extraordinary activity that can help people reduce stress levels and frustrations. I've been able to assist quite a few people with serious depression through using fur and feathers. For me personally, it has been an enormous help to de-stress myself; it kept my head clear and sharp at all times. Flytying was essential for me throughout my army work, especially during particularly tricky and high-risk responsibilities in my past. Flytying also helped quite a number of my army colleagues on missions or when they returned home again, to prevent or at least alleviate such problems as post-traumatic stress disorder. This is why I am still sending a lot of flytying materials to soldiers on missions all over the world and have sponsored organisations such as Healing Waters who are doing a great job for military personnel with problems.

Flytying makes you calmer, more creative and of course, it helps you to design and develop flies that you cannot buy in tackle shops. Flytying allows you to improve existing flies, or build more durable flies or even flies that become unsinkable. It is a great hobby for winter or long periods of bad weather. Catching your first fish with your own tied fly can even change your life. Flytying can be a passion or even an obsession; it can be art and it widens your knowledge about insect life enormously.

Beside marking the 40th (ruby!) anniversary of my Klinkhåmer, the motivation for this book comes purely out of passion and dedication to the sport of flyfishing, photography, and tying flies. This is why I picked a very *useful* selection of my best dry fly patterns which I use all over the world and over many years. I wanted to tell the real story of how my flies were conceived and developed and came to life. How and why I created my flies, improved myself as a flytyer and perfected my flies over nearly 50 years of flytying.

This book is a personal 'thank you' to all those friends who helped me to become a better flyfisher and flytyer. Of course, it is also a unique way to thank my sponsors who make it possible for me to hold so many free flytying workshops and classes for youngsters over the last 40 years, and into the future.

Thanks a million to all of you and happy tying!
Hans van Klinken
April 2024
Harskamp, Netherlands

Chapter 1

Dutch Panama
The forgotten fly

When I look back to the beginning of my flytying career, I always tell people that I started tying my own flies in 1976. That is largely true, but the fact is that I'd actually begun tying flies in 1974 and sadly gave it up rather quickly. I was eager to tie my own patterns because good flies were very expensive and often hard to find. In every tackle shop I visited in Scandinavia between 1971 and 1976 I bought an assortment of flies, until I started tying my own. It was a kind of private collection of local flies that I built up from all the different areas I fished in Scandinavia during my first years of flyfishing.

I was also fortunate in that so many flyfishers I met in Norway, Sweden or Finland gave me selections of their favourite flies. These are the two main reasons why my fly boxes were always well filled, and why I love to give away so many of my own flies to other anglers, once my tying had got up to standard. When I needed flies desperately, people gave me some or helped me out, but now that I have so many flies tied by myself, I can freely share them with other anglers. It feels great and it always brings back the good old memories of so many strangers that I met out in the wilds.

Fly boxes

I remember my first fly boxes very well. Two of them were just ordinary plastic compartment boxes that I had bought specially to store different sizes of dry flies. The third box was a classic aluminium Wheatley box with sixteen dry fly compartments on one side and a kind of foam on the other. I still have this fly box today, and it is not only my oldest fly box, but also the best quality aluminium dry fly box in my entire collection. I have replaced the foam a few times but what is remarkable is that none of the compartment doors has ever let me down. I used a lot of Wheatley boxes in the past and nowadays I use waterproof C&F fly boxes and I love them. The C&F are really worth having after you have dried your flies too many times when they fall into the water yet again.

My salmon and seatrout dry fly box was an old celluloid eight compartment fly box also made by Wheatley. Sadly I lost this box in 1988 while fishing the River Nordelva in Norway. The two Wheatleys I had bought second-hand, together with a very nice Fenwick glass fibre fly rod from a nice guy in Amsterdam who quit his flyfishing due to medical reasons. The aluminium Wheatley box even contained some flies and the name of the previous owner was scratched onto it, and is still visible today. I had only these four fly boxes while travelling through Scandinavia because my first flyfishing vest was completely handmade and only had space for four fly boxes. I have regrettably forgotten who made the vest for me, but I have worn it for years.

One plastic fly box I had was filled with size #12-10 dry flies for trout and grayling fishing in Scandinavia. My aluminium Wheatley box contained the same flies but in smaller sizes, #16-14, which I also used in Holland when fishing

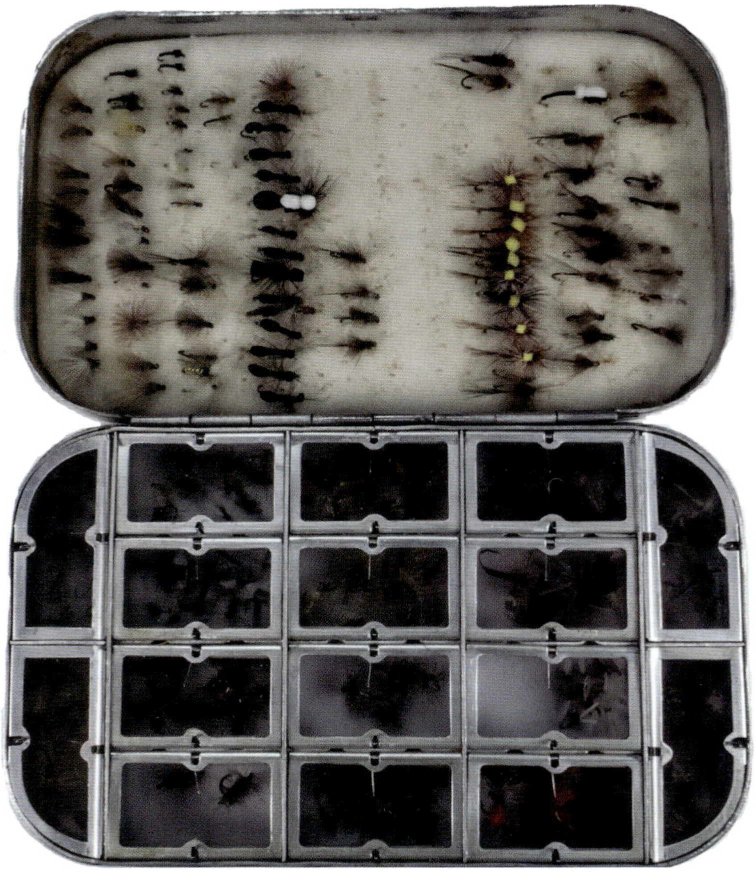

My first and oldest Wheatley fly box dated 1971

for rudd, roach and ide (orfe). My Wheatley celluloid fly box was filled with size 8-6 dry flies for salmon, seatrout and Arctic char, and my second plastic box holds all the flies that floated badly, or were damaged, or were patterns that were just never successful for me.

First diaries

I tried to tie my first flies in 1974 but I failed completely, and started afresh three times. Sadly there is no written record available of what exactly I did in the 1970s, because I only began making flytying and fishing notes in 1980 and only began writing my fishing diaries in the summer of 1984. That is the year when I designed my Klinkhåmer Special, which meant a record of the history of the development of my best pattern had been kept in the most comprehensive detail. I started writing my own diaries after I had seen examples of Arild Blomfeldt's brilliant artwork, a Scandinavian angler who decorated his diaries with beautiful aquarelles painted during his fishing trips. A few years later I met Arild again on the River Surna in Norway, and he allowed me to take a few pictures of the artworks in his private diaries. He even allowed me to use these pictures in a fishing story I was writing about Scandinavia. This book is a good opportunity to show a few more of his magnificent drawings to the rest of the world. As an expression of our friendship he did a lovely little grayling drawing for me before we said goodbye. I lost contact in 1989, but to me he was a real master at painting the pools, the landscapes, the flora and fauna, the fish he was catching and the flies he was using.

I was so impressed with his diaries that I simply had to start writing up my own notes more seriously, and today I am very lucky that I was so conscientious about keeping my own diaries. For most of my articles, book contributions and parts of this book, my diaries were an indispensable resource. However, most of what you read in this first chapter, I have written from memory, which was quite a tough job.

Just a selection from the 53 diaries that Ina and I wrote since 1984

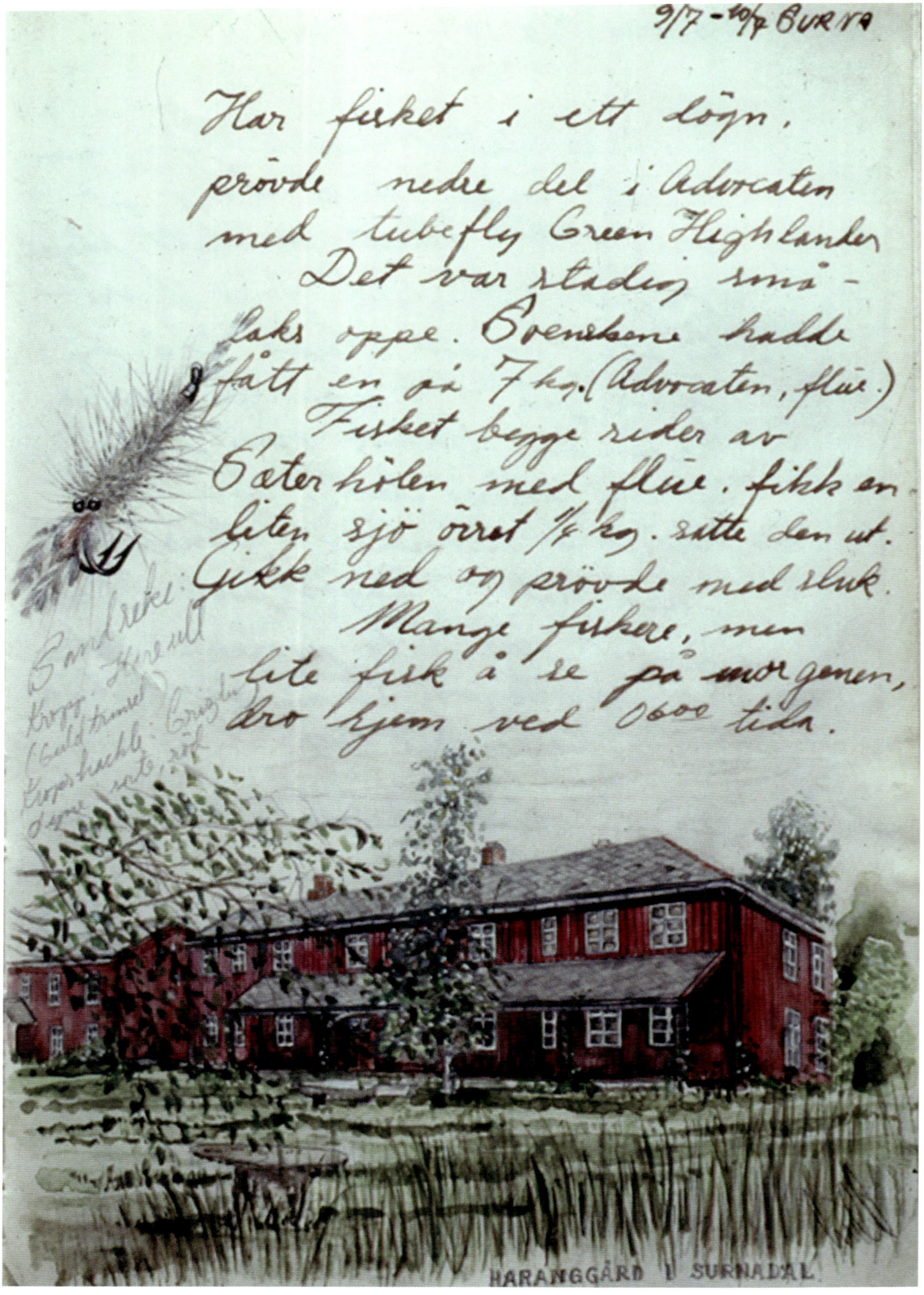
Har fisket i ett døgn.
prövde nedre del i Advocaten
med tubefly Green Highlander
Det var stadig some-
laks oppe. Svenskene hadde
fått en på 7 kg. (Advocaten, flue.)
Fisket begge sider av
Søterhölen med flue. fikk en
liten sjö örret ¼ kg. satte den ut.
Gikk ned og prövde med sluk.
Mange fiskere, men
lite fisk å se på morgenen,
dro hjem ved 0600 tida.

One of Arild's pages from his diary from the summer of 1984 that inspired me to start writing my own diaries
(© Arild Blomfeldt)

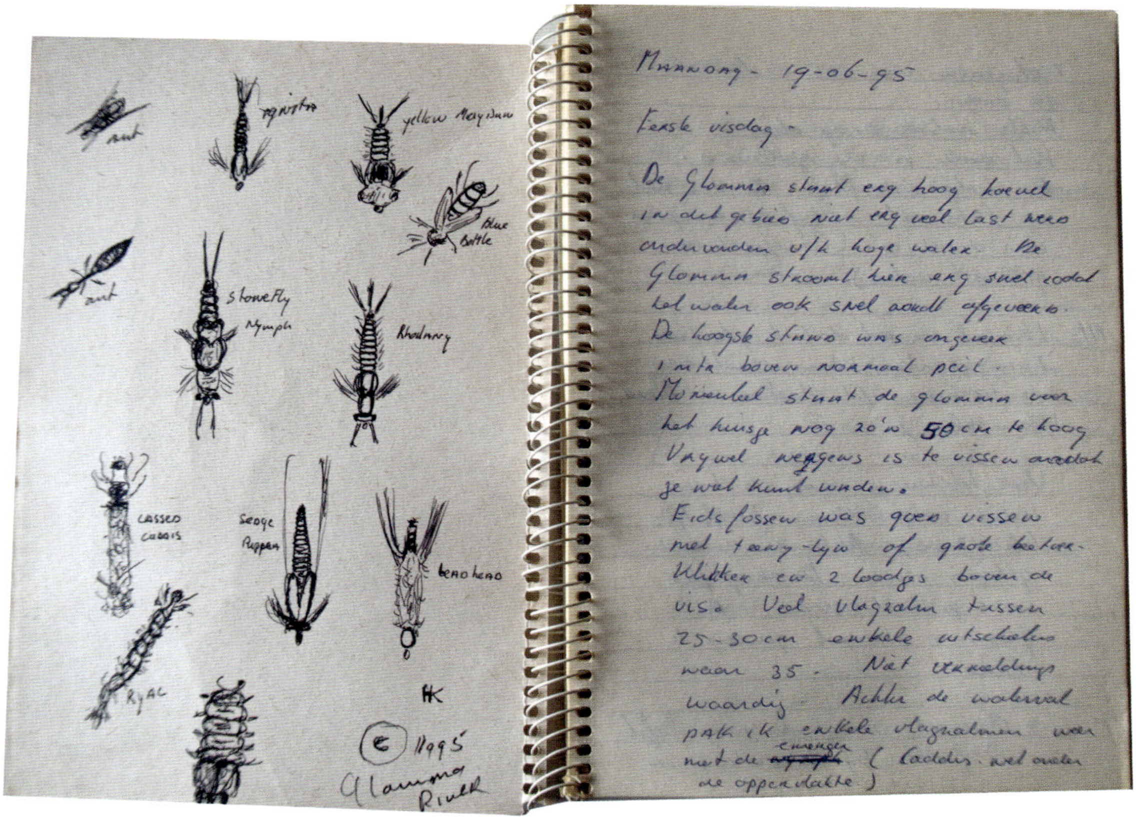

The inside of my 1995 diary. From 1996 I started to write all my diaries in English language

Start of flytying

In the early months of 1976 I finally began to understand some of the basic flytying techniques. This happened after I laid my hands on a copy of a little book called *Fly Tying* by Raymond Sugg, Ken Whitehead and Alan Vare. I got the book from the owner of my favourite tackle shop in Eindhoven, who had found it somewhere in the UK, and bought it for me for my twentieth birthday in January 1976. In those days there were hardly any flytying books available, especially where I lived in the south of Holland, and that little book I liked a lot.

I combined the techniques that I learned from the book with dissecting my most successful patterns with an old razor blade just to find out how they were put together. Two other very important flytying books in my learning process were, *The Complete Book of Fly Tying* by Eric Leiser (1977) and *Das grosse Buch vom Fliegenbinden* by Dr Klaus v Bredow (1981) written in German. Eric Leiser's book I loved so much that it became my personal flytying bible for years. So this is how I taught myself flytying. Without a teacher or Mr Google it was not easy, and good materials, books and quality tools were hard to find in the 70s and early 80s.

After a year playing around with thread, fur and feather I was captured by the art of flytying and was tying flies almost every day. Tying nymphs, streamers or wet flies was not an option because I was a real dry fly fanatic and purist for the first 10 years of my flyfishing life.

Of course, like so many other people, I started out tying simple dry flies such as a Red Tag, Bivisible, March Brown, Pale Olive Dun, Autumn Dun, Light and Black Cahill, Light and Dark Hendrickson, Quill Gordon, Black Spinner, Large Dark Spinner, Tup's Indispensable, Coch-y-Bonddu and all kinds of Palmers, to name but a few.

Most of the dry flies that I tied had no wings, and therefore are maybe better described as shoulder-hackled flies or collar-hackled flies.

Patterns with only a hackle in front were often described as collar-hackled flies and the area just behind the head of a fly was named the shoulder, which I believe is how the name for shoulder-hackled dry flies was also coined. So I learned both names a long time back.

I had added incentive to tie my own flies when my tackle dealer stopped selling flytying materials in December 1977. Much of it he gave to me during the first months of 1978, because I was one of his most loyal customers. By the middle of the winter of 1977-1978, a lot of new tools, threads, furs, feathers and a few amazingly good flytying books were suddenly increasingly available, and my motivation for flytying increased even further.

From then on I worked through Eric Leiser's book very seriously and my flies soon started to look better and better. I was also now focused on perfecting my tails, on making ribbing wraps more even and my bodies smoother. I practised a lot and slowly got more and more self-confident, becoming something of a fanatic and a far more creative tyer.

I had just had started as a sergeant in the Dutch army fulfilling my obligatory military service and was stationed in Germany (the Cold War was still on). If I was not on exercises or military duty, I didn't waste much time sitting in bars, but instead learned how to improve my flies or read some of the old classic tying books given to me. The money that I saved that way, I added to my fishing trips, or used it to buy yet more flytying materials.

Dutch Panama (version 1 and 2)

When I got a copy of *English Trout Flies* by W H Lawrie, my flytying life suddenly became a lot easier. It was like a huge eureka moment for me. Of course, nowadays this pattern book is indispensable when doing research, but in the late 1970s I mainly used it to find even more dry fly patterns without wings that I could tie. Although this book may have represented a boring collection of dressings to some people, I really like it. While I had no idea how most of the flies actually looked, it was still a great help in widening my knowledge and increasing my inspiration. One of these 'inspirations' was to create a picture in my mind of what a particular fly actually looked like when I only had the dressings and nothing else to go by. Of course it was impossible to tie the fly exactly as its creator had intended, but Lawrie's book showed me the importance of well-described pattern dressings, and it taught me to be as detailed as possible in recording my own dressings and tying descriptions from then on.

I don't make much of a secret of it, but in my young years I didn't like dry flies with wings made from feathers or quills. For me finding matching left and right wing segments was a tiresome problem, and anyway, the wings got damaged quite fast. For winging material for my own dry flies, I mainly used a bunch of white calf tail hair, or a solid clump of mallard fibres that I just divided in two equal wing parts with figure or eight and reverse figure of eight wraps. Sometimes I purposely left the entire bunch as one big dense wing, because in Scandinavia I didn't notice much difference in the catches I made with flies tied with a single wing or split wing patterns. Calf tail was inclined to get waterlogged after taking several fish, and is quite bulky when you tie it in, which is why, thanks to my favourite tackle dealer in Norway, I went over to using synthetic polypropylene yarn for my wings at an early stage in my flytying career.

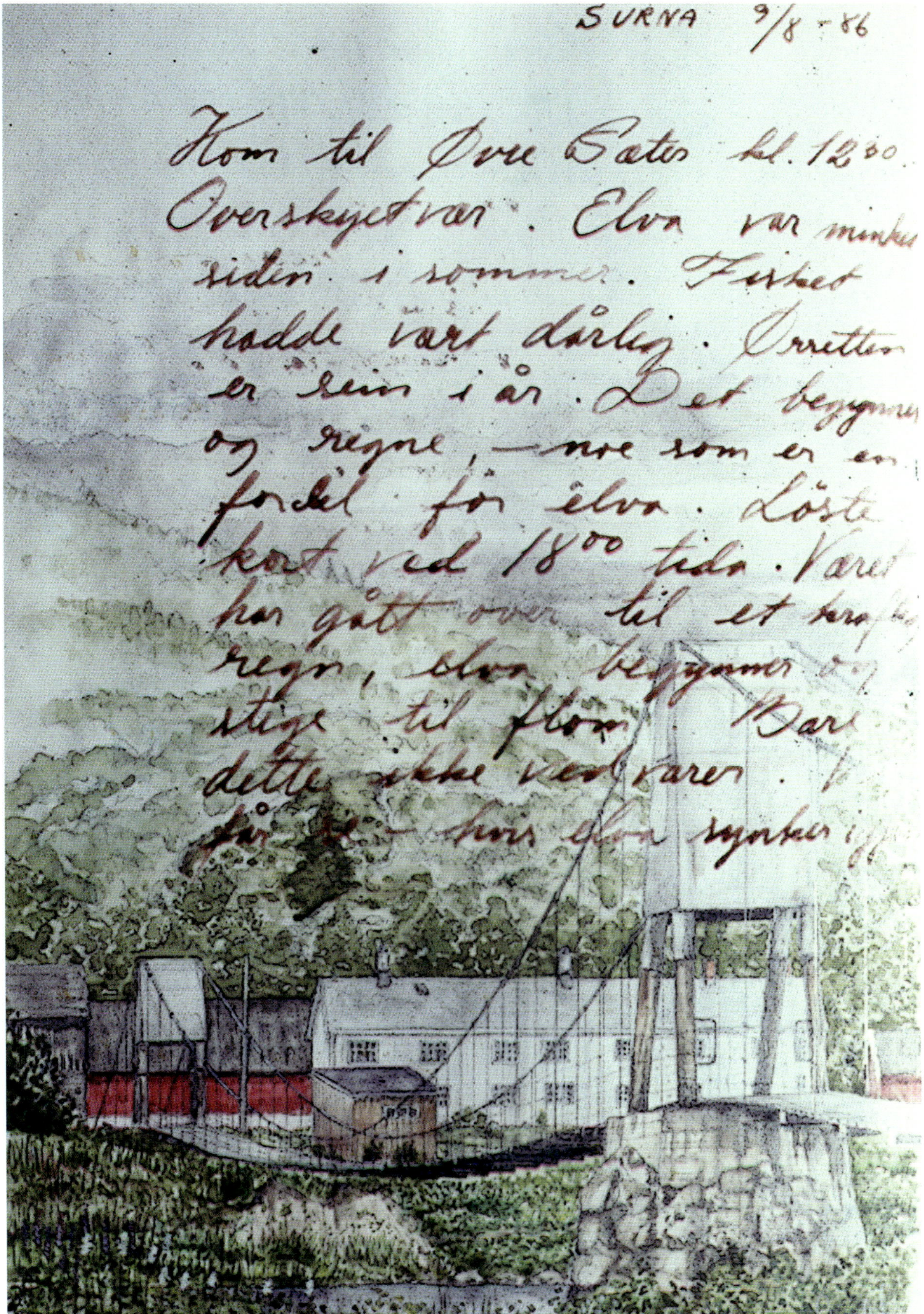

One of Arild's pages that he made in 1986 on the banks of the wonderful Surna river in Norway (© *Arild Blomfeldt*)

Scandinavian influence

Let's go back to my early days again. As soon it turned out that my girlfriend loved the outdoors, wildlife and wilderness as much as I did, I knew right away that I would spend most of my future summer holidays in the wild areas of Scandinavia. Knowing my girlfriend was happy to be outdoors meant I could concentrate fully on my flytying and making flies for Norway, Sweden and Finland. I started to change a lot of my patterns into more visible and better floating ones. I remember well how much effort I put into trying to improve the floatability of my flies. I also made bigger flies for my salmon fishing not only by using larger hooks, but also by increasing the number of hackle feathers to tie the collar, or by using longer hackles so I could add more wrappings. I also made bodies a lot thicker by creating egg shapes, or by experimenting with deer hair, and I tested many different materials to make more suitable tails. I often skipped the vital proportions and tried to do my flytying much more intuitively and in ways that allowed my flies to float and perform better in the turbulent and rugged rivers of Scandinavia.

I was very lucky that I had kept all my old patterns from previous years, even those that had been all but completely destroyed. I used the damaged flies as models for new designs and because only a little material was left on most of these hooks, I ended up developing a lot of fancy, even strange, patterns that certainly didn't make it in the end. But I never regarded that as wasted time because it helped me to improve my tying skills. Trying to rebuild or copy my old damaged flies, I often had to guess what materials had been used, and all this guessing had a great effect on the development of the flytyer I am today.

I also soon realised that I had to understand everything about good and useful tying materials and especially about how different materials reacted to water and how durable they really were while fishing. What might appear to be a good-looking fly at first sight could be a real disaster as soon as it was on the water. One of my top priorities in flytying became the durability of my flies because I was totally frustrated at the speed with which so many commercially tied flies fell apart.

Each fly that I finished I dropped into Permaflote (the best floatant in those days), and as soon as they dried out I tested them in a dish of water. I used all sizes of hackle material and even cut the hackle fibres into shape, but learned quickly enough that cutting hackle tips made flies get waterlogged much faster. I tested a lot of my flies from late April until the end of May in Dutch waters on rudd and ide. That was still the closed fishing season but I cut the bend off the hooks so I wasn't actually fishing in the real sense. When living in Germany I was able to test my dry flies in private army waters on salmonids, but I learned rapidly that fishing for stocked rainbows or brown trout was a lot different to serious river fishing in Scandinavia.

Most of my early patterns were copies of existing flies or flies that were given to me. That's why I have always claimed that my own first and serious fly development started after I had lost all my Rackelhanen patterns. It's also why my initial plan was to start this book with the Rackelhanen, because I hold the Rackelhanen, and the fact that I ran out of them, as responsible for my own particular style of flytying, and my further development as well. However, an important email I received a few years ago changed my mind. The sender asked me if I had any additional background information about a fly called the Dutch Panama. The man had browsed the World Wide Web and discovered that I had developed that fly. He gave me the link where I found a picture and once I saw the fly my memory flooded back and I realised that this was a totally forgotten pattern I'd tied, but definitely worth writing about again because I had caught countless fish on them in the late 70s!

The Dutch Panama I developed in the winter of 1977-1978 when I was trying to duplicate one of the Panama flies that I got from a very friendly Frenchman whom I had met in Sweden during my summer holiday in 1976. I loved this Panama pattern especially after making some extraordinary catches with it at several places that I fished when travelling to the northernmost part of Norway in 1976. I know I caught salmon and seatrout on it, but I am not sure how many. Once home, my tackle dealer with connections in France was able to get me a dozen Panamas, but they were unbelievably expensive, probably because of the complexity of the dressing. The only option left was to tie them myself. Sadly, I had destroyed the last original French Panama by taking it apart with the intention of copying it as best I could.

The original Panama is an ingenious pattern with four wings and three hackles, and as a novice flytyer it was so hard to tie that I decided to make a greatly simplified version by just making it wingless, with a body and shoulder hackle only. I actually made two versions as I had two versions of the original French pattern, perhaps representing a male and female. My first own Panama I tied with just one body hackle, a front hackle, and a tail from a few golden pheasant tippets. The second version I tied with a tail of fibres from a pheasant tail feather, ribbed the body with silk and a hackle, and used two different hackles in front. Both flies turned out a little different but I named them both the 'Dutch Panama', as they were still copies of the French patterns. I fished them very successfully for rudd, roach and ide in Dutch waters, and I caught lots of grayling in Sweden, Finland and Norway. I lost track of my Dutch Panama (and many other of my earliest shoulder hackled flies) once I started to use parachute flies in the early 80s which, as it turned out, were far more effective because they prevented grayling with their unique feeding style from pushing the fly away in the take.

Dutch Panama (version 1)

Tying the Dutch Panama (version 1)

Dressing

Hook:	Daiichi 1180 or 1182 size 10-16
Thread:	Semperfli Classic Waxed Thread 8/0, black
Tail:	Golden pheasant tippets
Rear body:	Peacock herl
Ribbing:	Natural red cock hackle over the front body
Front body:	Yellow or light yellow floss
Front hackle:	Natural reddish-brown cock hackle

Note: If you take a very high quality and long hackle you can skip the front hackle by using the same one as for the ribbing like I do. You also can use Semperfli black Straggle String as substitute for the peacock butt or rear body.

Tying Steps

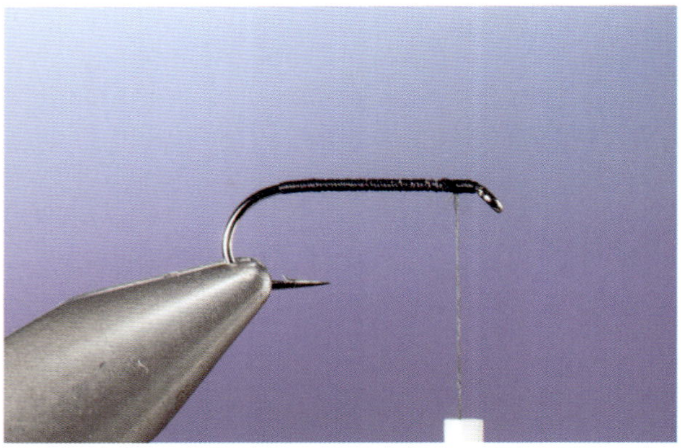

1 Place hook in the vice and wrap thread onto the hook shank from about 2mm behind the eye to the start of the hook bend and wrap back about ⅓ of the hook shank.

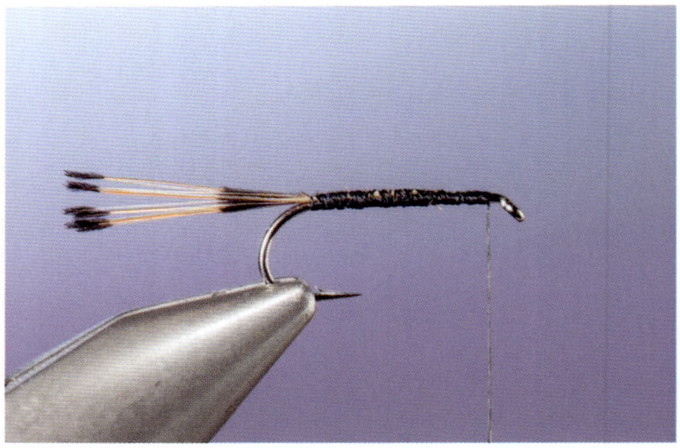

2 Select a nice bunch of gold pheasant tippet fibres and tie them in as tail. The tail length should be minimal equal to that of the hook shank.

3 Pick a high quality strand of peacock herl and secure it on the hook shank.

4 Wrap the peacock around the tying thread and wrap the herl forward in touching turns until ⅓ of hook shank is covered.

5 Select a nice, long, good quality hackle. Pull off the fibres at one side and tie in your body hackle at its tip.

6 Tie in your floss. I like most classic floss but also Semperfli yellow or cream colours.

7 Wrap a nice body with your floss. Cut off surplus and be sure that you leave at least 2mm free from the eye of the hook.

8 Now wrap the body hackle forward as ribbing and make three or four extra turns in front, secure hackle, and cut off.

9 Whip finish and varnish.

10 Completed fly.

Fishing in the Dolomites in Italy while trying the Dutch Panama (version 2)

Dutch Panama (version 2)

Beautiful Bosnia trout caught with the Dutch Panama (version 1)

Tying the Dutch Panama (version 2)

Dressing

Hook:	Daiichi 1280 size 10-16
Thread:	Semperfli classic waxed thread 8/0, black
Tail:	A little bunch of pheasant tail fibres
Tag:	Semperfli black floss
Ribbing:	Semperfli black floss
Body:	Natural raffia
Body hackle:	Brown cock hackle
Shoulder hackle:	Natural reddish-brown cock hackle
Front hackle:	A longer partridge feather picked from the side or mallard hackle as substitute

Tying Steps

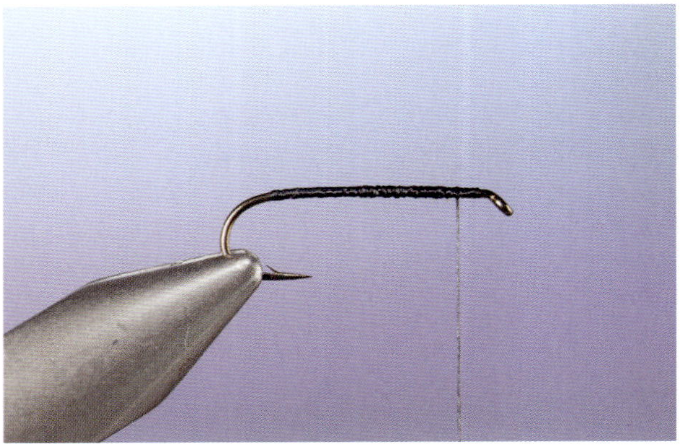

1 Place hook in the vice and wrap thread onto the hook shank from about 3mm behind the eye to the start of the hook bend and back again.

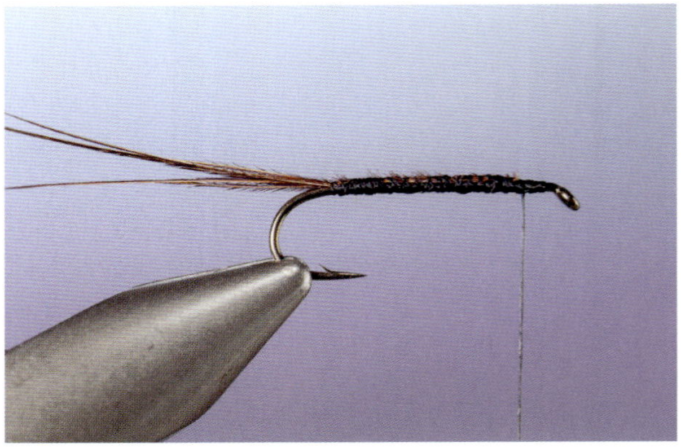

2 Cut out six to eight fibres from a male pheasant tail feather and tie them in at the tail end. The tail length should be at least equal to the length of the hook shank.

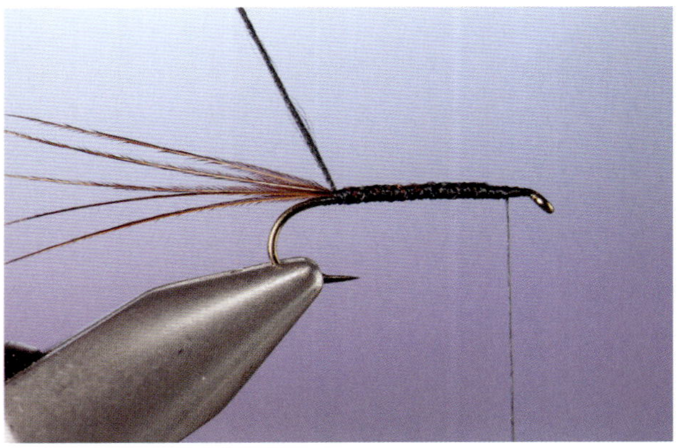

3 Tie in a piece of black floss silk.

4 Wrap your butt or tag about 3mm wide. Don't cut the silk as you will also need it for ribbing.

5 Select a nice long hackle, pull off the fibres from one side, and tie it in as your body hackle at its tip.

5a Take a small strip from your natural raffia. Don't use synthetic! I store it between the pages of a book. It will split easily.

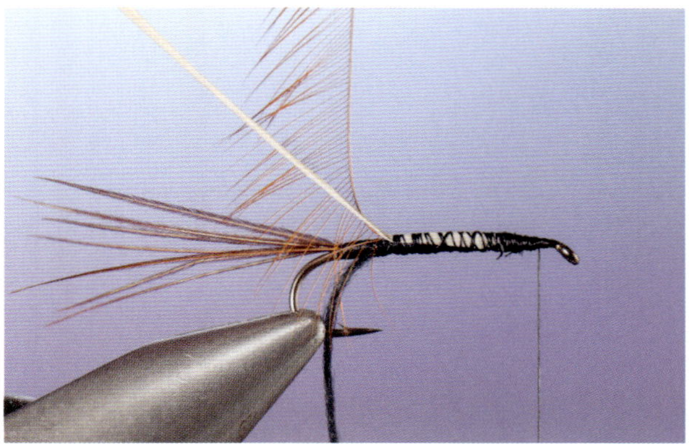

6 Wet the natural raffia first before you start working with it, and then tie it in.

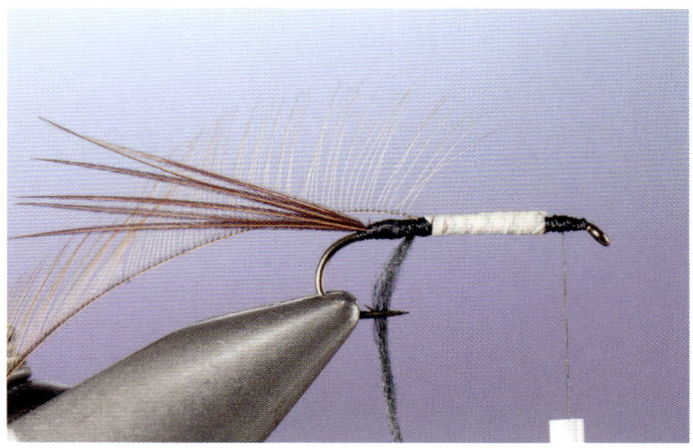

7 Wrap a nice body with the raffia. Cut off surplus and be sure that you leave at least 3mm free from the eye of the hook.

8 Wrap your floss as ribbing, secure and cut off.

9 Wrap the body hackle forward by letting it follow the black silk ribbing, secure and cut off surplus.

10 Tie in a reddish-brown shoulder hackle with fibres a little longer as the body hackle.

11 Wrap the shoulder hackle until you have a filled-up collar, secure and tie off but don't cut your thread. (Keep space in front)

12 Tie in the front hackle of partridge or mallard.

13 Make two full turns, secure, and cut off surplus.

14 Whip finish and varnish.

15 Completed fly.

I can't finish this chapter without including a reasonable copy of the original French Panama

CHAPTER 2

Rackelhanen

The big change in my flyfishing and flytying came in 1980 when I spent my summer holiday near the Swedish river Klarälven (Klara) with my girlfriend. The river runs through the province of Värmland, flowing from the Norwegian mountains southwards to feed Lake Vänern. Värmland is known throughout Sweden for its beautiful landscapes and cultural treasures. My plan was to start our journey in the little village of Sysslebäck and work our way up north as far as possible by following the river course.

An important meeting

After several days we came across an old angler close to the town of Kärrbackstrand who was having much more success than I was. Realising that this must be a really good fishing spot, we put up our tent and watched the man for quite some time. I was so interested in his skills and good results that I struck up a conversation with him on the riverbank. This particular meeting turned out to be one of the most important lessons in my flyfishing development. The man was very generous and willingly gave me useful tips and advice about the river and the area. The encounter became even more amazing when he started to explain his theories on flyfishing and how he related fishing with nature. Impressed by his stories and experiences, we offered him a nice shore lunch and glass of wine at our little campsite.

During his interesting discussion about flyfishing and his obvious respect for nature and wildlife, he showed me his fly boxes and shared with me the various fishing techniques he used. I was quite surprised that most of his patterns were mainly made from synthetics. I had never seen this kind of pattern before. He explained how he tied the flies, told me how to fish them and how to prepare them to keep them floating. I gave him a handful of my favourite flies and when he had to go home, he asked me to follow him to his car. He opened the boot, rummaged in some bags and gave me a little bottle of Permaflote, a few cards filled with poly yarn, a little box with hooks, and a dozen strange yellow and brown flies that he called 'Rackelhanen'. It was his favourite traditional Swedish sedge pattern and he emphatically insisted that I try it, using some of the techniques he had just explained to me. Then we said goodbye, and to show him how thankful I was, I gave him a send-off until his car had completely disappeared over the horizon.

Close to nature

Although my total flytying experience at that time amounted to just four years, my imitation of the old English standard hackled flies was not too bad, and tying them satisfied me very much. I also had no reason to develop other patterns, because I had no complaints with the catches I was making thus far. Still, this Rackelhanen interested me greatly and using it I had nearly as much success as my new-found Swedish friend had enjoyed. In no time, the Rackelhanen took its place among my favourite flies.

Full of confidence, I slowly travelled further up the route north along the river. We drove and walked many miles, camping out at the best fishing spots while enjoying the long dazzling midsummer nights. We lost all feeling for time and date and just took our naps when we felt tired and ate when we got hungry. For my girlfriend it was an unbelievable experience because it was her first real encounter with the untamed wilderness and she loved it.

The fishing knowledge I had gained on the small coastal streams above the Arctic Circle was a great help, but it was quite inadequate when it came to dealing with this huge, imposing river system. I soon began to appreciate the tips I had gleaned from the old man. While fishing I remembered his wise words; the silence and peace beside the river brought me very close to nature and wildlife, and I started to appreciate and respect it in a new way. On days when my catches were poor, or the weather was bad, I studied the varied and abundant insect species, and in time, I learned to appreciate them as well.

Like everybody else, I enjoy catching fish, and it gives me enormous satisfaction to watch a fish coming up to take my fly on the surface. But as time went by I realised that flyfishing was much more than just catching fish or casting a fly line. I soon discovered that catching and playing fish is very important in building up your skills and creating confidence and motivation. I also became aware that many other things are essential to becoming a proficient flyfisher. Today I am not a purist anymore, nor am I competitive and I like to observe the techniques others use in catching their fish. It is also really pleasant to share your knowledge and experience with others, and to share in their joy and satisfaction as well. Back then I also learned a lot from my girlfriend, who showed me how happy she was when I caught a nice size fish. It is all part of flyfishing, just as is watching animals at the waterside, which can also make my day.

I was learning quickly and the further north we went, the more connected I felt to this river system and its surroundings. With time, I also

Playing a marble trout caught with a Rackelhanen in the Idrijca river in Slovenia

Fishing for big Atlantic salmon with Rackelhanen patterns in Newfoundland and Labrador

learned how to improve my skills at catching larger fish. Occasionally, try as I might, I was unable to reach the location of a feeding fish. Instead of getting frustrated, I just took a longer break! I would sit on a nice boulder and eat my lunch while I gazed at the river. I watched rising fish for hours and it gave me new inspiration and many new ideas. I started to recognise holding places, food seams and feeding lies and I learned how to read a river. My casting in those days was not very good, neither in distance, nor technique. This caused me to avoid the slow water where the fishing challenge was much higher. My poor distance-casting quickly made the fast broken water and rapids my favourite spots.

Completely new style

After fishing at several locations and being happy with my catches, my luck changed for the worse. I lost my last new 'favourite fly', and had to revert to my old English standard-hackled dry flies again. They were still very effective, especially the Red Tag and Greenwell's Glory, but the Rackelhanen had achieved so much that I had no other choice but to make a copy. After trying several designs, using the old man's yarn and hooks, I finally arrived at my own version of the Rackelhanen, which, despite some minor tying mistakes and flotation problems, seemed to be a reasonable copy of the original.

I can only guess how different my flyfishing life would be today without this wonderful encounter on the riverbank and the discovery and opportunity of trying the Rackelhanen. The success of this fly gave me enormous self-confidence when tying my own patterns and it started me off on a completely new style of flytying. With my own variation of the Rackelhanen, which I called the Poly Sedge, my creative flytying development began. For me it's simple: without the Rackelhanen there never would have been a Klinkhåmer Special, or the Once and Away or the Parapoly Sedge in my fly boxes today.

Rackelhanen background

The Rackelhanen is a Swedish name that implies a fly that is a cross between a male Black Grouse (Lyrurus tetrix) and female Western Capercaillie (Tetrao urogallus). Both have recently been moved to their own family of Tetraonidae. In my opinion, the Rackelhanen is much more than just an ordinary fly. I have enormous respect for the originator, and it is a pity that only a small group of flyfishers know the thinking behind the development of this wonderful pattern. It is undoubtedly one of the most simple fly patterns I know. It is so easy to tie that it is the perfect beginner's pattern. The Rackelhanen is also an ideal pattern for experimentation. It is a highly underutilised fly, unknown beyond Scandinavia. It is masterfully designed and I have never seen a pattern that imitates the pupa, adult and spent caddis better than the Rackelhanen.

Kenneth Boström, from Helsingborg, developed this very popular Nordic pattern in 1967. It came during one of his fishing trips to the River Vännån, on the borders of the Swedish provinces Kronoberg and Halland. In the 1960s, this river was one of Sweden's best in the

Kenneth Boström 1945-2020 (© Kenneth Boström)

Two original Rackehanen specially tied for this chapter by Kenneth Boström

Two of Kenneth's flies with his signature

south. Kenneth told me he observed spectacular hatches of many types of caddis and upwinged flies, and found the number and variety of insect species in this river to be incredible.

His idea for the Rackelhanen came to him during a late evening hatch of caddis. The trout were being very selective, but were most definitely interested in the ascending pupae. The pupae were being taken just under the surface. Adults on the surface were disregarded, and flew away safely. This particular feeding behaviour occurred most evenings and for days on end. This allowed Kenneth the time needed to find the most effective method to construct a fly which would imitate the ascending pupae.

His choice of the right combination of material (polypropylene), a special tying technique, and a unique fishing method, made his pattern deadly. Unfortunately, the completed fly looked ugly and therefore it was a long time before Kenneth had the courage to show his creation to his close friends. Sadly, the Rackelhanen has never truly received the recognition it deserves, in part because of its simplicity, its name, and its appearance. But all of us who have fished with it quickly came to appreciate in the span of just a few casts that it will out-fish most of the more famous and traditional patterns.

We are fortunate to enjoy over 200 caddis species in Scandinavia. This makes the Rackelhanen not only a very effective fly pattern in Sweden, but all over Northern Europe as well. The fly hangs deeply but naturally in the surface film. This is most important for grayling feeding selectively, or for finicky trout under certain circumstances. The fly produces a beautiful surface wake upon retrieval, very similar to that produced when flies are tied with the famous Newfoundland or riffling hitch. The appearance of the Rackelhanen, when observed against the sky from below, forms a nice fuzzy silhouette. The poly yarn allows for a very lifelike impression, because caddis behaviour is very different to that of an upwinged fly.

Rackelhanen fishing techniques

Even if you tie an exact copy of this pattern, you will not succeed without using the correct fishing techniques. The most productive technique I use to imitate the ascending pupae is to fish the fly downstream. Draw it under the surface film with short but powerful pulls, and then allow it to rise again. When the fly is drawn under the surface, it will capture air bubbles and cause a very life-like imitation of the pupa. To imitate the spent caddis, you simply split the wings before casting.

Kenneth described several of his techniques for the rackelhanen. Because his creation is mainly fished in or on the surface, he prefers a floating line but in some circumstances, a sink tip line can be highly recommended.

In both situations, the fly must always be prepared with floatant!

His most effective fishing methods are:

Dragging it on the surface

Kenneth uses short, jerky pulls, or pulls about 5cm long, with a short pause between them. This imitates the natural behaviour of the egg-laying caddis, or a drinking caddis.

When he sees the fish eating

'Swimming', 'paddling', or 'struggling' caddis pupae on the water, he usually cuts off the wings of the Rackelhanen and presents it as a surface-swimming pupa that is heading for land.

One of thousands of beautiful rivers that look perfect for using a Rackelhanen fly

A lake outflow is always a hot spot for the Rackelhanen

Colourful release after it took a Rackelhanen

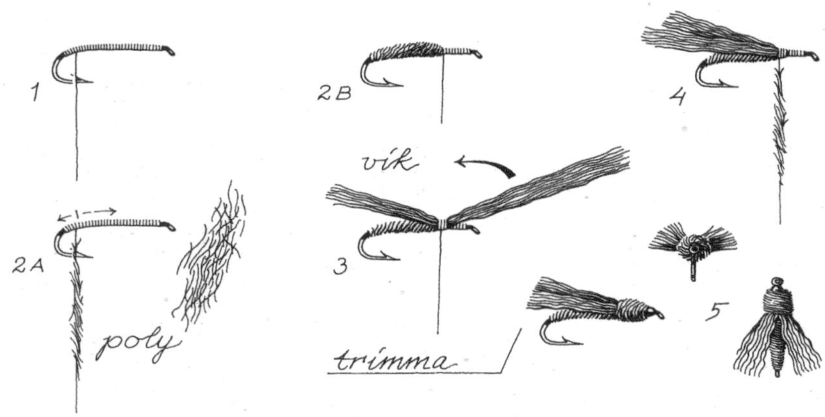

1. Jag använder helst färdigvaxad tråd, och avstämmer denna till den färg jag tänker ge min färdiga fluga. Jag lägger en bädd med bindtråden över krokskaftet ner mot böjen, och vänder därefter och lindar några få millimeter tillbaka.

Kenneth's famous little booklet that he published about his Rackelhanen (© Kenneth Boström)

Sink-tip method

During all the years I fished in Scandinavia, I only once saw another flyfisher using this technique, but I use it a lot in Norway, Sweden and Canada. Use a sink-tip line or sinking leader with a short tippet between 80-120cm in length. Lower your rod tip and pull the line so that the fly drags under the water surface. Make a short pause, and the fly will start to float up again. This technique gives a perfect imitation of a caddis that is hatching just below the surface. Personally, I quite often use the same technique when fishing for pike with large floating lures.

Dead drift method

With this method, the Rackelhanen floats drag-free on surface. This fishing technique imitates a dead caddis after egg laying.

Nice grilse took a Rackelhanen with white wing and brown body

Original Rackelhanen tied by Kenneth and given to me at one of the flyfishing fairs in Sweden

Tying the Rackelhanen

To tie the original Rackelhanen, you need only a hook, a good wax, tying thread and poly yarn. The poly yarn should always be treated with a floating substance to make it resistant to absorption. In spite of the fact that poly yarn has a specific gravity of 0.94, it readily absorbs water and the fly will ultimately sink. Many people have misunderstood this quality of polypropylene, and believe that it floats well without treatment. Trust me, it won't, because the weight of the hook will pull the fly down! To give the fly its best floatability, we used Permaflote in the old days as it was highly recommended by the originator. It is a great floatant indeed, with the only disadvantage being that it can be quite caustic and dissolve the varnish that I use to protect the knots of my flies. To solve that problem I have used Dilly Wax exclusively since the late 1980s.

Dressing

Hook:	Daiichi 1280, 1180 or 1182 in sizes 8-10 and 12
Thread:	Semperfli classic waxed thread 6/0, dark brown or black
Body:	Poly yarn in your preferred colour (match the hatch)
Wings:	Poly yarn in the same colour as body
Thorax:	Poly yarn in the same colour as body

Note: Extra tools needed: Dubbing wax, dubbing twister (I prefer electric one) coffee grinder (optional). Test your poly yarn by cutting off a piece, make a little pellet and dropping it in water. Check particularly for two things: it must absorb quickly and stay in the surface. If it sinks, it's NOT poly yarn. If you use Semperfli's poly yarn, use double strands because it's thinner than most other brands.

Safety note: Never use super glue in combination with poly yarn because it causes a dangerous chemical reaction and can burn the synthetic fibres into your skin. It also can damage your health when breathing the chemical gases.

Tying steps

1 Select your favourite colour of poly yarn to match the hatch. Of course you also can mix colours.

2 Cut off about 50 little pieces of yarn approximately 1 cm long. For this particular fly I have chosen to make my dubbing from brown yarn.

3 Put the pieces in an electric coffee grinder and activate the machine by pressing the power switch on for 1 second. Repeat until all the little pieces have changed into a nice bunch of dubbing.

4 Never pulse the power switch longer than one second because the yarn fibres will get too hot and will merge together or burn and the dubbing will be useless.

5 If you have no grinder, don't worry because you still can make your dubbing by rubbing the little pieces between both hands or between your hand and a piece of hard foam.

6 If you are used to a grinder, you can mix all kinds of materials, even synthetic and natural. I love to work with poly yarn because it will blend different colours so nicely.

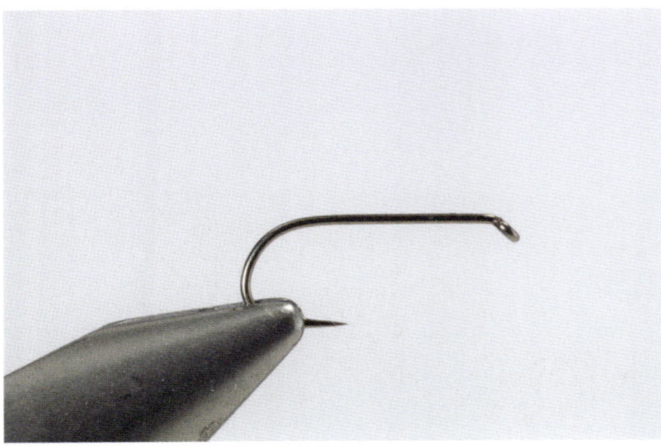

7 Now that the most important preparations are done, place your hook in the vice.

8 Cover the entire hook shank with your tying thread. Leave your thread close to the bend.

9 Put some high-quality wax on your tying thread. Believe me, it will make the dubbing process a lot easier, especially when working with dubbing made from rough poly yarn.

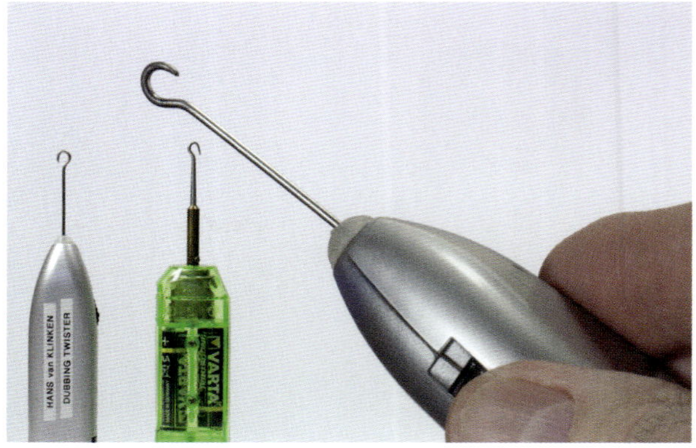

10 A few of my self-made electronic dubbing twisters that I bought for less than one euro. One was once a mini fan, the other one a milk frother. All I did was make a nice tapered point on them and reshaped it.

11 Spread out the shaggy dubbing over the tying thread and twist the yarn around your thread between forefinger and thumb. Then make your loop!

12 Now it is time to grab your dubbing tool or use an electric dubbing twister like I do. I can highly recommend the use of an electronic twister. You will be amazed with the result.

13 Make the first two wraps to the hook bend. Then dub back in the opposite direction towards the eye, making sure the ragged fibres at the end are well secured underneath.

14 Now form a nice tapered body. Leave some extra space for the wings.

15 Cut a piece of poly yarn about 6cm long.

16 Tie in the yarn in the middle, just in front of the body. This step of Kenneth's increases the durability of the fly.

17 By tying each wing separately, it becomes impossible to pull out the wing when the fly is completed. It is a simple but clever idea.

18 Be sure one end extends over the eye of the hook. Once the yarn is secured well, twist the wing and position it slightly left on the body.

19 Now continue by pulling the other piece of yarn into position, making it the right wing. Be sure to lock both wings securely with your tying thread.

20 By using this technique, you make the most durable wing imaginable. In this step it is also very important to produce a smooth underbody for the thorax.

21 Put some extra wax or a drop of varnish on your underbody. This may give you some extra grip when you dub the thorax.

22 Set up a new dubbing loop that is filled with some extra dubbing.

23 Wrap a neat thorax but do this very carefully and slowly. When finished, form a nice head with wraps of the tying thread.

24 Wrap a nice head and tie off by using your whip finishing tool or any other technique that you prefer.

25 Trim the yarn wing fibres just beyond the hook bend as shown in the picture above.

26 Your Rackelhanen is finished.

27 and 28 In these pictures it is clear that when seen from above the Rackelhar en has two wings.

My Poly Sedge (below left) and Spinnertail (below right) which I used for two years

Chapter 3

The Way with a Parachute

Although my own variations of the Swedish Rackelhanen, my Poly Sedges and my Spinnertails were all pretty good patterns during my fishing season in 1981, it sometimes appeared as if there was something wrong with them. Under exactly the same conditions, fanatically rising fish sometimes seemed to refuse the pattern completely, while other fish took it immediately and with a lot of aggression. In other situations, I felt that I missed a good fish that I normally should have hooked quite easily. It did not happen every cast, but enough to cause me some serious doubts about the effectiveness of these two copies of the Rackelhanen that I had created myself. This also is the reason why I haven't put their tying steps in this book and why I haven't written about them in the past more often. It was a real mystery because I did not have these problems with the original Rackelhanen.

It posed many riddles and punished my brain for quite a while. I had no explanation for it, and it took me some time to discover the problem. As usual, I found the reason by accident and I remember it very well. I had produced a nice series of Poly Sedges and Spinnertails for the upcoming season in December 1981. I had tied them on various types of hooks, on different hook sizes and with dissimilar wing lengths, just to be sure I was well prepared for all kinds of unexpected fishing conditions. I also had used new body and wing colours, with the intention of doing some extra experiments with colours for the summer of 1982 in Norway and Sweden. I had prepared all these flies with Permaflote as soon they were finished and when dried, I put them in my fly box. I learned that trick by bad experience after I had lost my bottle of floatant twice in previous years. When dropping the flies in the liquid after they left the vice, some of the chemicals stayed behind and I could start fishing with them with optimum floatability, and no further treatment was needed before fishing.

Then, a few weeks later in early February 1982, I dropped some of my newly tied Poly Sedges and Spinnertails in a dish of water to show their floatability to a friend. During this demonstration, I saw, with great astonishment, how one of the patterns floated in what I call 'a wing-flat position'. This means that the fly is lying or floating on its side instead, with the hook point hanging underneath. I didn't like it and it forced me to do two more tests with and without floatant.

I made another handful of brand new Poly Sedges and Spinnertails that I didn't prepare with any floatant at all and dropped them in my dish of water. All these patterns floated perfectly in the surface film, but only for a short moment because as soon the poly yarn body and wing absorbed the water, the weight of the hook was enough to let them sink. This proved what I already knew, which is that a good floatant was indispensable to make poly yarn water-resistant and keep my Poly flies floating.

The second test I did with patterns in which only the wing was prepared with floatant. It was

On the right my Spinnertail floating perfectly, with wing up, on the left my Poly Sedge in a wing-flat position

a lot harder to do this because you mainly use Permaflote to dunk the whole fly in the little bottle, shake it and that's all. Most of the flies with wing-only floatant didn't have any problems floating, and they all drifted perfectly but now and then a fly was floating in wing-flat position. I was very sure that the wing-flat position was the main cause for the problems with my Poly Sedges and Spinnertails. This problem was extremely difficult to recognise while fishing in the broken surface water of the mighty rivers like the Glomma or Klara. For me, it became an important explanation why fish sometimes refused to rise, and why I failed to hook fish that should have been caught easily. The problem was fully clear to me after I had done my tests, because when flies are not floating properly, the silhouette of the pattern, as it should be seen from below, is totally missing.

The only reason why I believe some fish still took it was because they probably saw the fly as a kind of terrestrial. Another lesson I clearly learned was that when preparing the wing of a fly only with floatant, it gives the fly a better and more natural drift, with the body hanging a little deeper in the surface film.

Important note: *At this stage of my flytying I didn't know the designer of the Rackelhanen and I also wasn't familiar with his thoughts behind his fly. I was just trying to solve a problem that was actually a serious tying mistake but I simply didn't know it. How wrong I was, I only realised when I got some original Rackelhanen flies from Kenneth Boström in Sweden a few years later.*

I then saw my mistake right away because all his patterns were tied with two wings instead of one wing as I had done with my Poly Sedges and Spinnertails. Of course, I did my tests again with Kenneth's flies to be 100% sure and none of them floated in a wing-flat position. Kenneth was also using different hooks and his dubbed bodies were a little bulkier too, which probably reduced the problem as well.

My mistake had a good side too because it was the start of my making parachute flies and it had pushed me accidentally into the most important stage of my flytying career.

The discovery of a second and even third problem

In the spring of 1982, I still tied and prepared a lot of standard hackled dry flies for the upcoming season and I made them with nice solid tails, long wings and good size hackles. Thus, I could see my flies much better because I was able to concentrate on the drift of my flies a lot easier. I love to watch how trout and grayling come up for artificial flies and real insects. After several years tying these standard dry flies, they also started to look quite nice as well. In the late 1970s and early 1980s a lot of my conventional dry flies were tied in same way as master flytyer Eric Leiser did in his *Complete Book of Fly Tying*. The book was printed in 1977 and it surely was my flytying bible for a very long time. I loved the way Eric taught his readers how to tie the wings of the Light Cahill.

He wrote: *As with most dry flies you tie in the wings of the Light Cahill first. That's to your advantage, since the wings are the most difficult to master. If you don't tie them correctly, you can start over again at this point instead of having to dismantle the entire pattern.*

In June 1982, when fishing the Secret river in eastern Norway, and presenting a well hackled Wickham's Fancy, I made another remarkable discovery. I had tied the fly with a bunch of wood duck fibres as wings instead using an original left

My conventional dry flies at the end of the 1970s. I mostly tied them with mallard fibres with one or two wings

and right mallard wing quill. In simple terms, I can call it a Wickham's Fancy built with wings in Light Cahill style. My great luck in all this was the short distance presentation in a deep but narrow channel in the middle of the river. The channel itself was perhaps only three metres wide while the entire width of the river was nearly thirty metres. To reach the best spot I had to wade closer to the edge of the channel and present my flies on rather short casts. This awesome place is one of my best hotspots for fishing in Scandinavia because the channel continues for about 800 metres and you could easily fish it for a couple of hours. I mostly fish it in the morning, after lunch or after dinner so only four hours per day to prevent it getting overfished. Again, I proved to myself how important good observation really is.

Today I see what I observed then as the most important and powerful of my entire flyfishing career. It was the 'Lady of the Stream' – the grayling – that brought another even more serious problem to my attention.

Not more than three metres in front of me, I saw clearly and with great astonishment, how my fly was pushed aside as a grayling aggressively tried to take it from the surface. It was not taken at all because the grayling had clearly missed the fly during her speedy rise. Upon closer observation, I saw a few more times how the heavy hackled fly was pushed either upward or to one side. The channel was packed with fish which made it a perfect testing area. Today I know that most of those fish were migrating grayling returning to the much bigger Glomma river from their spawning grounds almost 28 kilometres away.

I kept casting and presenting my fly to these fanatical and very hungry feeding fish and made a third important discovery. I realised that when you see fish coming up so quickly, you set the hook pretty fast too, actually much too fast. When doing lots of classes and workshops many years later, I discovered this happens to a lot people because they are focussing on the upcoming fish more than the fly. You really have to learn to keep concentrating on the *fly* at all times. Too fast reflexes snatch the fly away before the fish can take it. A full second delay helps a bit, and prevents people for missing a few good fish but my most important lessons learned

The top of the Secret river in high water at the end of May

Although we catch a lot grayling with dry flies, her protruding upper lip makes her a perfect bottom feeder

that particular day was that as long we used the traditional standard and heavy hackled dry flies, the problem still exists while fishing for grayling. Any delay in setting the hook will not work in that case. Being too fast mainly occurs when fishing dry flies too close to the fish and when you see the fish coming. It's a serious problem with salmon fishing with dry flies, which I first discovered only many years later.

Trying to understand why standard hackled flies are often pushed away or to one side, I followed this line of thinking: grayling can come up at very high speed to take flies from the surface film, and I don't know 100% sure, but I think that it is probably a combination of the shape of her mouth, the speed of rise, and the way of taking the fly that creates a tiny pressure ridge in front of the grayling's mouth. This pressure is what I think pushes the fly sideways or forward and the bushier the flies are hackled, the more often they were accidentally pushed away. We all know that because of the grayling's protruding upper lip, she is a perfect bottom feeder, but that doesn't mean that she isn't able to feed on dry flies, emergers or terrestrials at the surface. I had many great discussions on this subject with flytyer Oliver Edwards and we always agreed about this amazing feeding ability of the grayling. Sadly, Oliver passed away in April 2023 and I will really miss him.

Over the years I learned that grayling can feed from the surface extremely well but you need a really clever fly to catch every rise that you notice. You also must be able to recognise the grayling rise as well so you can plan your casting strategy, especially in the small streams of Denmark. Grayling have adapted themselves perfectly for taking insects from the surface film. How fast grayling come up we discovered firstly in 2004 after getting home from a long study on grayling in Canada. Many studies went well but while checking ten hours of professional underwater footage of grayling taking dry flies on the Wolf river in Yukon Territory in Canada, we only clearly saw twice how a grayling took a fly from the surface with their belly up, while all the other shots were blurred or simply too fast to see anything clearly. The cameraman had put his settings at 24 shots a second and had checked his captures only in the viewfinder, thinking they all were ok, but sadly they weren't.

My first ever parachute flies, The Parachute (left) and ParaDun (right)

My first parachute flies

My floatability problems with the Poly Sedges and Spinnertails, in combination with my observations about the grayling take, kept me busy and made me want to create a completely different pattern as soon as possible. A great idea finally arose at the end of June 1982 while still fishing in Norway, and I went through all the flytying books that I had with me on my holiday. I often read them while it was raining, when I was searching for something or I just was relaxing in our tent.

The answer was to add a **horizontal** hackle, and wrap it around a wing root as I also saw in *The Complete Book of Fly Tying*. Eric's parachute chapter was amazing, well described and richly filled with beautiful and clear drawings. If a parachute fly could be made around a hairy wing version of the Royal Coachman, why should I not be able to manage it with my Poly Sedges? Where Eric used calf tail hair, mallard feather fibres or hackle tips as wings, I used my poly yarn wing. I only had to tie the wing from my Poly Sedges a little further back, put it upright with my tying thread, and apply some extra dubbing in front of the wing. Then start wrapping the parachute hackle around the wing base I made at the bottom of my poly yarn wing. After a good study of the parachute chapter and a few attempts later, I succeeded, and my first serious parachute pattern was born. Wrapping a hackle around a wing of poly yarn had worked extremely well and I was not just very happy but unbelievably proud of my new creation. Immediately after the fly was ready, I dropped it in my dish of water. The floatability was far above all expectations, especially after I treated my first parachute flies with Permaflote. I called my first version 'The Parachute' and a few days later, when I made a variation with a tail, I called it 'The ParaDun'. Both were developed on the banks of the Glomma river.

Conclusion: beside Kennet Boström, Eric Leiser was also responsible for 'my way with Parachutes.' Once I developed my Klinkhåmer Special, I discovered that a deep-hanging surface film position reduces failed hooking almost to zero. Ever since I started using my 'iceberg fly' the Klinkhåmer, I have never seen the fly pushed aside again.

In the spell of the parachute fly

In no time parachute flies became my favourite flies for grayling fishing and once I had tied my first one, I started fishing them almost exclusively, with huge confidence and great success. I

My first variation of the Yellow May Fly that I created in the early 1980s

Two classic Greenwell's Glorys tied in parachute-style as sold by Alex Martin in the 1930s

Classic Tup's Indispensable as sold by Alex Martin in the 1930s

still use them a lot today and that's exactly 43 years in a row, which proves my addiction and preference for them. I am completely in the spell of tying and fishing parachute flies. Meanwhile there is so much to say on the subject in terms of my research and development as I discovered them. We can safely claim that the inventors of parachute flies gave dry fly fishing a whole new dimension and that they did flyfishers a great favour. Yet this special unusual dry fly concept did not immediately cause a revolution in the flyfishing world. Parachute flies gradually gained in popularity over the years. They have been tied and fished for nearly a century, which makes it difficult to unearth the complete early history of them.

A little history of parachute flies

Parachute flies are specially built with the hackle wound horizontally around an upright wing or projection, almost like an open umbrella. This can be on top, or even underneath, the hook shank, instead of vertically around the hook shank, as we would normally do in conventional or standard-hackled dry flies. Parachute flies are relatively new when compared to traditional standard-hackle flies, but their past is far more complicated, perhaps because not so many people were tying them. Good and reliable background information about parachute flies is extremely hard to find. To name who exactly designed the first parachute style of hackling is simply impossible. I had to give up my research numerous times. There are many contradictory assertions about dates, and potential first inventors. I had dozens of discussions with people who claimed to have some knowledge about parachute flies but often we ended up disagreeing. This made my research mainly a one-way investigation over a long period of time. I will never say I discovered the truth, not least because you never know if somewhere on the globe, somebody had that similar idea too. In the past there was no internet, databases or social

Partridge & Yellow parachute as sold by Alex Martin

media as we have it today. However, what I found and discovered about parachute flies is without doubt worth sharing with those who might be interested.

Many patterns and tying methods I researched from the past were described so briefly and vaguely that I simply had to guess how the hackle was wrapped because tying steps were hardly mentioned. Dates for the invention of the parachute style varied between 1912-1933. An English writer claimed in the 1980s that the parachute fly was invented around 1912 in Australia but he didn't elaborate.

Joseph Spear Beck wrote in his article *Parachute Pioneers* in the *American Fly Fisher* and published by the Museum of American Fly Fishing many decades ago: *that John Dickson & Son Ltd., in a note dated 1970, while crediting the parachute fly as an American invention, was of the opinion that it was first marketed about 1928 in the United States but that no company in the US thought it was worth pursuing.*

How wrong they were is history now. In the same article, I read that Alex Martin's company had merged with the firm of John Dickson & Son Ltd in the early 1960s.

Harold Hinsdill Smedley stated in his book *Fly Patterns and Their Origins (1946)* that the parachute fly was the invention of William A. Brush and refers to his granted US patent from 1934.

Others claim it was designed by the Scots then taken up by the Americans: I really didn't know what was true or not. However, quite a few stories mentioned that the Scots invented the parachute style of hackling. The only thing I know for sure is that several flies tied with a parachute hackle on top of the shank have been around since the early 1930s. It has often been claimed that Alex Martin, a Scottish tackle dealer from Glasgow, invented the parachute fly, but I think that those statements are mainly based on the fact that he was one of the biggest suppliers of parachute flies in the 1930s.

Two Parachutes that I made the classic way. Left I used a mast of biots and right I used a wild pig bristle

An old parachute fly I picked up in the UK, the hackle clearly wrapped around a mast of a pig bristle

Another old parachute fly somebody gave me, the mast clearly made from goose biots

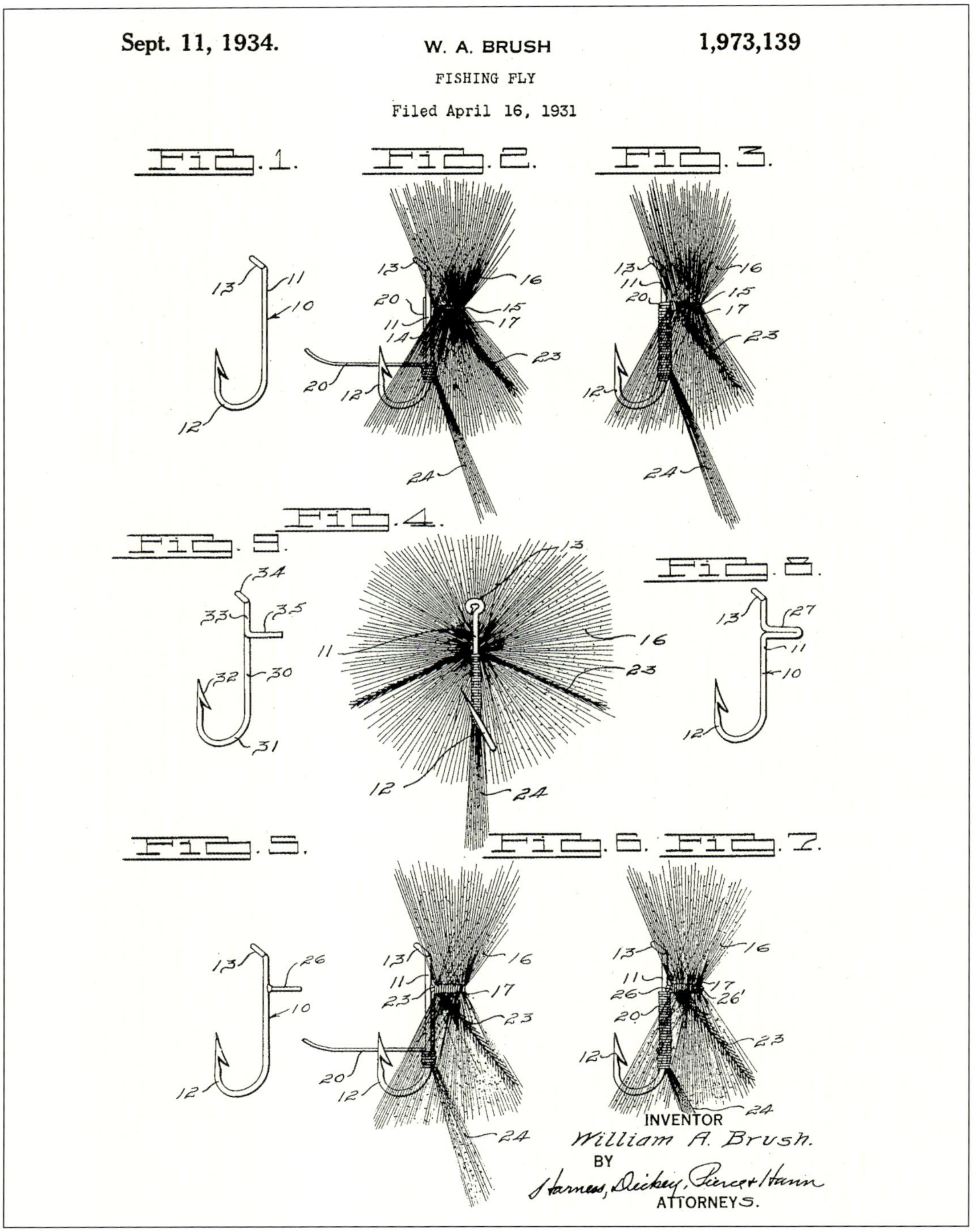

Brush US patent filed 16 April 1931 and first granted at 11 September 1934 with the US publication number 1973139A

Description of the drawings from Brush's patent.

Fig. 1 *is an elevation view of a conventional type of fish hook before a hackle is secured thereto;*
Fig 2 *illustrates the hook shown in fig. 1, and the manner of securing the hackle to it;*
Fig.3 *illustrates the construction shown by fig.2 after the hackle has finally been secured to the hook, or in other words this figure illustrates a completed artificial fly constructed according to one form of the invention.*
Fig. 4 *is a view of the construction shown by Fig.3 taken from the underside thereof;*
Fig 5 *illustrates a slightly different form of metal hook which may be used;*
Fig 6 *illustrates the manner of securing the hackle to the hook shank shown in fig.5;*
Fig 7 *illustrates the construction shown by fig 6 after the hackle has finally been secured to the hook;*
Fig 8 *illustrates another form of metal hook which may be used;*
Fig 9 *illustrates still another form of metal hook which may be used.*

John Crossman amazed the fishing world with his excellent article about Helen Todd in *Trout and Salmon* magazine in 1981, and he cleared up a lot of confusion about the Alex Martin stories. Helen Todd, a wonderfully gifted fly dresser from Scotland, worked as a professional flytyer for Wallace & Kerr before she joined the firm of Alex Martin for an extra eight shillings a week. Today we know that she devised the parachute style of hackling in the winter of 1931-32. It also showed up in several stories that Alex Martin patented and marketed the parachute pattern in 1933, but I couldn't find any proof of a patent request by Alex Martin in my research. Helen got her inspiration after reading an article in an American angling magazine about a fly with the hackle fibres tied in a spent-wing position. She wound the hackle horizontally around a mast from gut or even a pig's bristle, which was secured on the top of the hook shank. I also saw several old parachute flies with a mast or stem from stiff quill fibres and when experimenting myself I found goose biots absolutely awesome for the trick. Many people still see Helen as the true inventor of the parachute fly today.

I respect all tyers, especially these from the past and I believe that many of them came up with the same ideas, tricks or similar patterns without knowing it from each other. Exactly like it happened to me with my Klinkhåmer fly and the extra weight as I placed it on my Leadhead as well. Personally, I think the most prominent name associated with the parachute hackling style is William Avery Brush from Detroit, Michigan. Brush was an engineer and co-owner of Brush Motor Company, maker of the 'Brush Runabout' automobile. The Runabout is long gone, but Brush's parachute style in dry flies is still doing extremely well into the 21st century.

I was very fortunate to get an official copy of Brush's original UK patent request which was sent to me personally by James L Hardy. In this six-page patent specification, Brush describes **three** ingenious ideas of attaching or making a projection on the hook to which the hackle can be attached off-set from the shank. Two hooks have a kind of hackling spindle at the thorax, opposite to the bend, on which the hackle was wrapped. His fourth idea described the use of a feather 'quill' to wrap the hackle around. Brush filed his US patent on 16 April 1931 and it was first granted on 11 September 1934 with US publication number 1973139A.

Brush described the design in his patent specification as follows: '*an artificial fly in which the hackle is so related to the hook that when the latter is in its properly suspended position, the position of the fly will closely correspond to the position that a live fly would have on the water.*'

PATENT SPECIFICATION

379,343

Application Date: May 27, 1931. No. 15,534/31.

Complete Accepted: Aug. 29, 1932.

COMPLETE SPECIFICATION.

I, WILLIAM Avery Brush, a citizen of the United States of America, of 1913, Palmer Park Boulevard, Detroit, Michigan, United States of America, do hereby declare the nature of this invention and in what manner the same is to be performed, to be particularly described and ascertained in and by the following statement:

William Avery Brush British Patent no 379,343 published 29 August 1932

Brush British patent

Brush was a very clever man because he also applied for his English patent on 27 May 1931, even before Helen Todd had tied her first parachute. It got registration number 15,534/31. The patent was granted and published on 29 August 1932 and numbered 379,343 (officially GB379343A). This famous patent number showed up in many reports and articles about parachute flies later. I am very sure that Brush and his friends had already been experimenting for a while with his parachute flies long before 1931; who knows, possibly by the mid 1920s. I am very sure that he first applied for the patent only after he knew the trick with the parachute style would work. Brush's specification and the descriptions included in his patent request were amazing. He patented the improvements in or relating to artificial flies. The patent related to the projection on the hook and not to the fly itself. I don't know if the patents from Brush were applied to make some extra money but a few conjectures were made in a direction of commercial benefits.

Another patent (no. 619417) was taken out by E K Gregory in December 1946 with no. 36317 (class 48) and refers clearly to the Brush patent 379343. Gregory uses one hackle from which he bends away the stiff hackle quill and winds the more flexible part around it.

When doing further research I discovered that William Brush's parachute design was initially tied and sold commercially as the 'Gyrofly' by the well-respected William Mills and Son Company on Park Place in New York City. It was one of the longest-running sporting goods stores in the USA. The company was started in 1822 under the name T & J Bate and changed name five times until it finally became Wm Mills & Son in 1875. The name 'Gyrofly' for the parachute design was invented by A C Mills Sr and was used exclusively by them. Perhaps this name popped up because the hackle was horizontally placed on top of the fly like the rotors on a helicopter. At first the parachute fly was licensed exclusively to this firm by Brush. Ray Bergman, who worked for them as a salesman, confirms this in his excellent book *Trout* (1966), where he illustrates an example of an early parachute fly, a Ginger Quill, with hackle-tip wings. At the time, a very popular fly was the Parachute Adams. Later, more companies were licensed by Brush to sell the parachute flies.

Hardy Brothers enter the scene

In England, the Hardy Brothers of Alnwick marketed them as 'Ride-Rite' flies. Hardy's 'Ride-Rite' dry flies, tied on the up-eye 'Ride-Rite' hook were very revolutionary in those days, but in reality, they were tied on one of William Brush's hook inventions. The hackles of this series of flies were wrapped around a projection on top of the hook shank, with the idea of supporting the fly in the correct position on the surface film, giving the fly a better representation of the natural fly. In January 1937, a royalty of 1/- (5p) was paid by Hardy's to Alex Martin for Ride-Rite flies dressed on streamline hooks. The flies were sold for 5/6d per dozen, which made the royalty 18.5%. I have no idea what connection Alex Martin had with Rite-Rite flies or streamline hooks. The Rite-Rite hooks were also made by Partridge of Redditch

619,417. Fishing. GREGORY, E. K. Dec. 9, 1946, No. 36317. [Class 48]

An artificial dry fly the hackle of which is secured to the side of the shank A remote from the hook comprises a feather tied at a part A^2 intermediate its length to the shank with the stiffer part B bent away from the shank and the more flexible part B^4 wound about the part B. The hackle is trimmed to the shape shown in Fig. 5. A body, tail whisk and wings are added. Specification 379,343 is referred to.

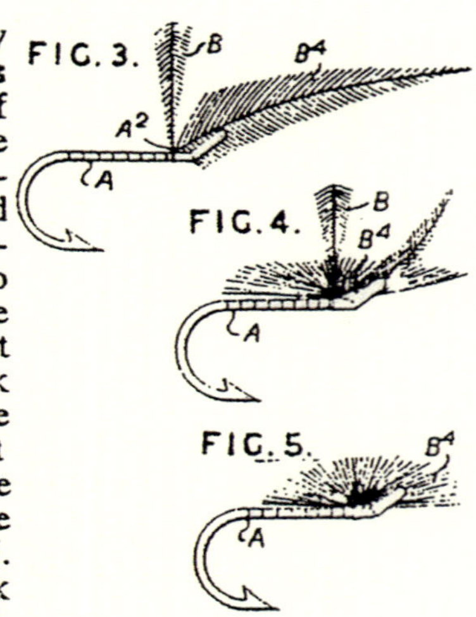

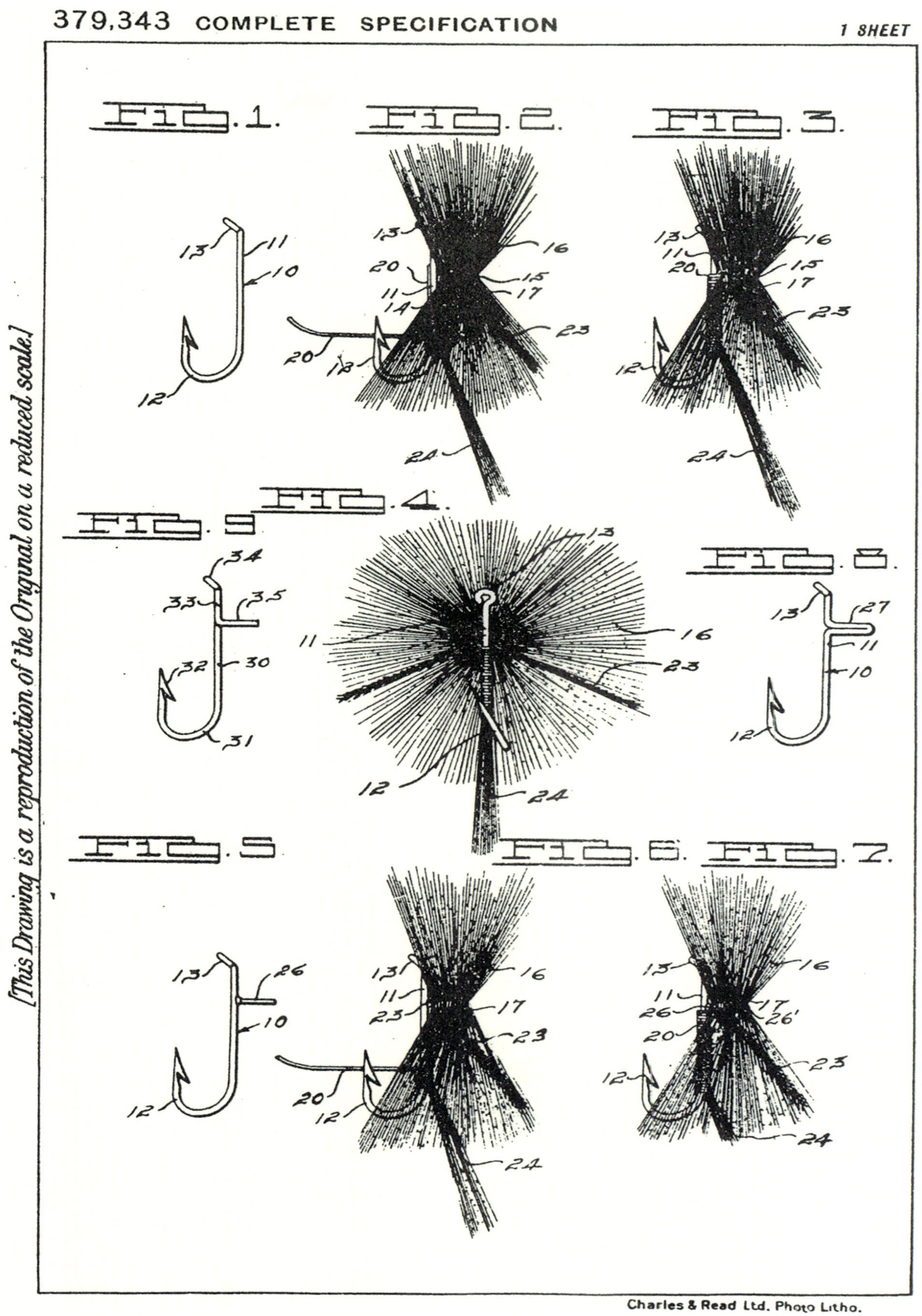

Brush UK Patent 379,343 given to me by James L. Hardy

Alex Martin outpoint hooks

Alex Martin's invisible silver hooks

but James Hardy told me that there will not be any old Ride-Rite hooks left at Hardy's as all obsolete hooks were thrown out when the factory moved in December 1964.

The Alex Martin product catalogue of 1937 shows that the firm marketed the parachute flies in the same year and from that moment on started improving them. Martin's firm sold parachute flies for trout, seatrout, salmon and also a special mayfly series. According to the catalogue all these flies were good for fishing dry as well as wet. They were sold at home and abroad and were well received.

Brush writes thus to Alex Martin around 1935: '*Your Parachute Flies are the most beautiful fabricated flies I have ever known.*' And later: '*I have complimented your organization before and can't refrain from doing so again. The Parachutes are perfectly beautiful things.*'

Martin used special hooks for his parachute flies which he called 'outpoint hooks'. These hooks had points that were somewhat off-set from the shank and were specially sharpened to ensure better hooking. For some of his flies he also used 'Martin's invisible silverhooks', which were forged steel outpoint hooks with fine silver plating that in no way blunted the hook and presented to the trout a less easily-seen surface. These hooks were used for fly production but they were also sold individually.

In my search, I also discovered another patented hook that was used for parachutes in the US and UK in the old days. (British Patent no. 438819) I don't know the exact date, but it must have been somewhere around 1937. All these patents made everything really confusing, at least for me. The hook was invented by Dr E R L Gow, and he used them to produce his special 'Aero' dry flies. The patent is for the hook in which he developed a small projection with a hook eye, which was situated on top of the shank, near the natural position of the head of the fly. On the size chart, I noticed that the three available sizes (0-3 or 16, 15, 14, 13) still had a normal up-eye as well. The hook was specially invented to keep the gut out of the water film, and not specifically for tying parachutes. 'Aero' flies dressed with spread wings were simply called 'Aero', and those with upright wings were called 'Aero Upright'. The Dr Gow hook was entirely handmade in England by Sealey's for Herter's Inc USA, and better known as the model 432 parachute dry fly hook. Most famous were sizes 8-10 and 12. Hardy also paid royalties to Dr Gow in the beginning of 1937. This royalty was 6d (2½p) for Streamline hooks and was about 9%. This must have all to do with the Ride-Rite flies that Hardy sold. Herter's produced two versions of the 432, the reason for which is unknown to me. The hook had a true 'Perfect Round Bend' and the 'Eyed Mast' was placed with a post at the halfway mark, while the other one had a post at the 2/3rd mark. The date of this hook would be sometime during the mid 1930s.

Other parachute hooks

Partridge also produced special hooks for parachute flies about 75 years ago but I don't have much information about them. But their latest hooks especially made for tying parachutes were the Partridge HE1 hooks produced in 1997.

The most recent patent related to parachute flies was requested by Lee Wulff on 6 August 1987 and was granted on 18 October 1988 with no. US4777759. I know about his patent and the related fly very well. It was a Surface Stone Fly that had a hook with a moulded plastic body which incorporated the parachute post. He gave me a few to try in Newfoundland and Labrador.

The Dr. Gow "Aero" Dry Flies

ON PATENT HOOKS

British Patent No. 438819

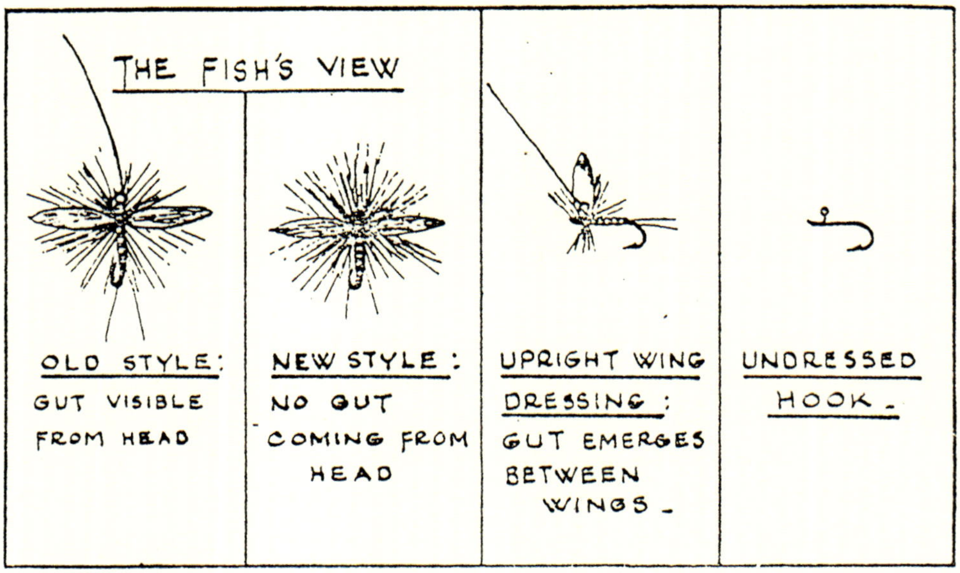

Hitherto in Dry Fly Fishing the gut attached to the head of the fly must necessarily either rest on the surface of the water or else be in the water. Both of these cases have the disadvantage that the gut is directly visible to the fish. With this new style of hook, the invention of Dr. E. R. L. Gow, this disadvantage is to a great extent eliminated. The feature of the hook is that the eye is situated on top of the shank, near to the natural position of the head of the fly.

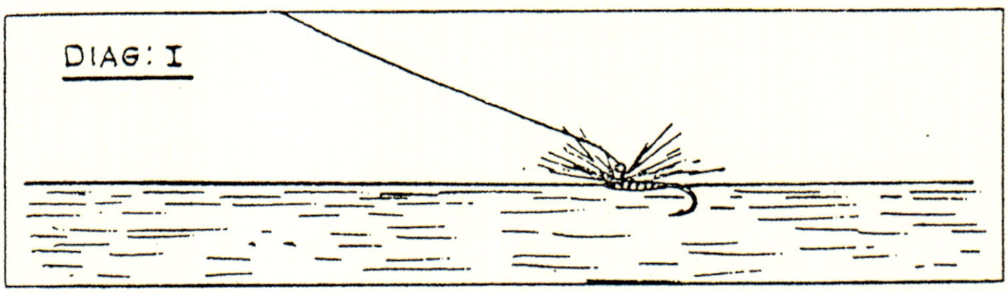

Diagram I shows the position of fly and gut at the instant that the fly alights on the water. There is no gut near the water and therefore the fly appears to the fish to be **completely natural.** This is a great advantage when casting over rising fish. The "Aero" Fly always lands as a natural fly because of its perfect balance.

It can be proved by the Principle of Refraction of Light that the gut in the air, when viewed by the fish from below the surface of the water, appears to be further away than it is in reality. To explain further—when we look **from air into water,** the bed of the river appears nearer to us than it really is, so when the fish look **from water to air** the process is reversed and objects in the air appear further away.

C4

Aero dry flies from Dr Gow

It was described as: *An artificial fishing fly having a body on the shank of the hook, there being a headed post rising from the body, to which post a parachute is tied in the form of a hackle feather wound on the post.*

Background of the invention regarding the official patent: *This invention relates generally to an artificial fishing fly referred to as a Parachute fly. This fly is generally fished dry and is characterised by a hackle wholly above the hook shank, i.e., opposite the hook, this hackle being wound and tied circularly with its fibres extending in a plane in all directions. Other, more conventional dressing may be added, but the Parachute fly imitates better, floats better on the water surface, and is more stable. Some fly casters are of the opinion that it casts better and comes to settle on the water in a more life-like manner, and floats on the water surface film, rather than above the surface, as do more conventional types of dry flies. However, the Parachute fly is more difficult to tie properly, and this invention provides for easier tying, while making the Parachute better and more uniform from fly to fly. It is important that the hackle should be spread uniformly in a circle and lie in a fixed plane neatly.*

Note: All the patents that I mention in this chapter have now expired!

Today there are many different parachute flies

Special tools, exciting new tying techniques, even hooks re-shaped, re-designed and improved, have evolved just to help people winding a parachute hackle more easily. Just think about the Swedish dry fly hook (1979), the revised 'Ride-Rite' hook, Bob Barlow's Gallows tool also known as the Stem-Loop Parachute method, popularised by John Veniard and Donald Downs in 1972. Then there's Tomas Olsson's 'heat-saddle', or melted parachute hackle technique (1983), and the clever idea from the late Staffan Lindstrøm using a tiny drop of melt glue to create his series of parachutes. Very effective are parachute Sedges (*see chapter about the Para Poly Sedge*). The Halo

Stunning upside down parachute fly tied by Andrea Malzanini

One of my large parachute Mayflies created by hackling around caribou or deer hair on a Daiichi 1230 hook

Benefits of using parachute flies as described in the early days

Mr Brush claimed that a parachute fly would float longer due to the horizontally-wound hackle providing more contact with the water surface thereby spreading the weight of the fly over a larger area making optimal use of the water's surface tension. He also claimed that parachute flies allow subtle presentation because they drop slowly and gently onto the water.
Alex Martin also explains the advantages of these new flies. He lists these in his catalogues:

1. That as a wet fly they swim right side up, as a dry fly they float right side up.
2. Oiling is dispensed with.
3. That they 'light' on the water instead of splashing onto it.
4. That the hackle points remain on the surface as do the legs of the natural insect.
5. That the whole length of the body is visible to the fish as in the natural fly.
6. That in shape and aspect, the natural is represented in a singularly accurate manner.
7. That no legs, hackle, or material shield the hook.
8. That a 'Parachute' can be fished as a floater or as a wet fly.
9. That the hook meets the fish.

Many agree that Martin's fifth-listed advantage is one of the most important. Yet William Brush didn't even mention it.

Extended body parachute Mayfly with calf tail wing

Balloon Parachutes tied by Branko Gašparin, a close friend and master flytyer from Slovenia

Large parachute sedges with wings from poly yarn and deer hair fished and tied by Branko Gašparin

CDC Puff Ball Parachute, designed by Pierre Mouchel in 2005-2006

Hackled Parachutes or Klinkhåmers, a mixture of my Klinkhåmer and the sequence of Morten Oland (Denmark) and the added halo hackle by Tom Sutcliffe (South Africa) are all great improvements. Flies like the Water Wisp which needs its own hook to secure it in the vice, and the Reverse Parachute Mayfly as tied by the late Roy Christie also give you a good idea how creative people can be with fur and feathers. Amazing also is the Tandem Parachute Fly that is tied with two wing posts and two hackles (a big one in front and smaller one at the back). Sometimes an Upside Down Parachute (tied on the bottom of the hook shank) can work excellently too. Worth to mention are also the parachute CDC flies with sequins from Gerd Peter Wiedits and Oliver Edwards' Mohican Mayfly.

A great parachute fly we can't ignore is the 'Branko Killer', a fly that is perfect for all the rivers in Slovenia. My friend Branko Gašparin, owner of the Villa Noblesa Lodge in Most na Soči, devised it years ago. He discovered that a simple parachute fly tied with lilac-rose materials is exceptionally effective. Branko loves to share his theories about the special effect of the lilac-rose colour – and his flies! – with all his visitors. Many of them love the Branko Killer and restrict themselves to this pattern. It is a fact – and I can confirm this – that Branko Killers seem to have an almost magical attraction for fish, especially for grayling. There are so many more parachute flies and I just mention a few.

The real power of the parachute fly

A lot of people still might believe that a conventional dry fly floats on the hackle tips and tail only, but that's *not* the case. Some of us might remember the famous catalogue from Dan Bailey's Fly Shop in Livingston, Montana where the renowned Royal Wulff fly was featured on the cover for many years. The fly was pictured daintily standing on what appeared to be water

Here I wrapped the hackle around a Wally Wing to create a Wally Wing Parachute fly. This beautiful wing technique was invented in the 1970s by Canadian flyfisher and tyer Wally Lutz

Reverse parachute Mayflies tied by Roy Christy

Beautiful extended body Mohican Mayfly tied by Oliver Edwards

Trends that have evolved in the last few decades

1. Parachute flies are no longer being tied on special hooks with masts, due to the advent of new materials in combination with improved tying techniques.
2. Special parachute hooks are no longer produced.
3. You seldom see parachute flies without post or wings. Most tyers like a visible post. It serves as a wing and is a lot easier to see. This is important during hatches and spinner falls, or when the fly is hard to spot in low light or bright sunlight.
4. Many tyers work with synthetic materials today, not exclusively for the body only, but also for the posts and tails. These materials are light, strong and available in many colours, densities and thicknesses.
5. Parachute flies are used for a wide variety of imitations nowadays: stoneflies, sedges, upwinged flies, ants etc.
6. They are also being tied on curved hooks to create my special iceberg effect, as I do with my Klinkhåmer Special for which I designed a very special hook indeed.

Gerd-Peter Wieditz CDC Parachute with a sequin on top to keep the parachute in better shape

on the tips of hackle barbs and tail. I always listen with fascination when my good friend Wayne Luallen explains to tyers at shows and seminars how misleading that really was. He tells them that the photo was probably taken with the fly sitting on gelatine! People laugh when he says so, but then he has proved it with a freshly tied Royal Wulff and a cup of water. He takes the fly and carefully drops it from about three centimetres off the water and it immediately

Tandem Parachute with 2 wings and 2 hackles. I used it for Atlantic salmon in the 1980-1990s

Alex Martin's Daddy Long Legs Parachute from the 1930s

This is a box containing three boxes of unused Alex Martin's flies now belonging to Leon Links. The outer box is $4\frac{3}{8}$ inches square (11cm) and the inner boxes are about 2 inches square (5cm). The previous owner told Leon that he ordered these just before the Second World War but only got to use one box before he was called up for service. I have always wondered what happened to the first owner in the war, and also to the fourth missing box and what flies were in it

Last year Leon got the opportunity to buy two bigger flies, a Parachute Mayfly pattern and a Parachute Daddy Long Legs. Both are from the Grass Body series, with detached bodies and tied on outpoint hooks. My 1937 Martin catalogue says of the Grass Bodies: 'A blade of dry grass floating on the stream floats unaffected by ruffled waters, through the rapids and ever on the surface like a live thing in its element'

These flies are from the 1930s and were sold in pretty carton boxes that each held a few samples of a particular fly pattern: three Partridge & Yellows (on outpoint hooks), two Tup's Indispensables (on silvered hooks), and two Greenwell's Glorys (on standard hooks). Little labels with Alex Martin's Patent No 378343 were attached to the hook points

Michael Olesen's powerful Luna Mayfly Parachute

Parachute Damsels given to me by A E Daffurn

sinks about halfway down to the body. I had a big laugh when I looked at their faces and saw how impressed they were. Then he takes the fly out and blows off all the water and drops it carefully again. This time it sinks all the way to the body, flush in the film, almost like a parachute fly.

I was unbelievably lucky to discover at an early stage of my flyfishing development, the importance and power of parachute flies. I also noticed that flyfishing with parachutes has many benefits.

My first benefit starts with the presentation. Because of the flat hackle in combination with a nice parachute cast, the fly falls gently on the surface in a much more natural way. In my personal opinion, it is one of the most stable dry flies, and with a good knowledge of flytying materials, you can use them in turbulent water and even short rapids without any problems. Parachute flies also give you a much better imitation of real insects, because they create a far better silhouette as well. The full circle of wrapped horizontal hackle fibres also increases the fly's floatability, at least when you tie fine delicate parachute flies. As a result, I no longer had to use as much floatant as before and used it on the poly yarn wing only.

My first parachute flies were quite heavily hackled, maybe too much, because I strongly believed the bigger the parachute hackle, the better the fly would float. It was a very wrong thought and I discovered not that much later that longer hackle fibres were a far better option and caught far more fish. When I also began to tie my flies a bit shaggier and purposely *uglier*, they started to work even better in the numerous fast-broken waters where I prefer to fish so eagerly. I also noticed that more voluminous patterns

Foam CDC parachutes as I tied them for the Guden river in Denmark

clearly brought up fewer fish when I tried them in quieter or even stillwaters. This was another lesson I had to learn, and I decided to test more patterns with fewer hackle windings at locations where the current was smoother and in stillwater. It was amazing to see that if a parachute fly is tied properly, you don't need an exaggerated hackle to keep the fly floating. After a little more experience, I even was able to fish with exactly the same flies in a much greater variety of water conditions and even on stillwaters.

When parachute flies became more popular, flytyers also began to discover other tying techniques and started to use different materials. Nowadays we see parachute flies with or without wings, with extended bodies, and with or without tails. We are using cock hackles, CDC, squirrel hairs, and there seem endless variations.

The parachute tying technique and the materials

There are several tying methods for developing a parachute fly, but I will limit myself to a few techniques only in this book. Not because other tricks are too complicated, or that I am too lazy, but simply because I have tied so many parachute flies using these techniques, and I feel most familiar with them. In most situations, it is the way of securing the hackle which differentiates one technique from another.

If you understand the tricks of hackling and how to tie off the hackle, a parachute fly is just as simple to tie as any other fly. To remove a lot of confusion, one must understand how a well-tied parachute fly floats on or in the surface film.

For small parachute flies the wing of a parachute fly should not support the pattern to float on the surface film. It is the *hackle* that keeps the fly afloat; sometimes supported by a tail. The fact that a parachute fly doesn't need a wing or tail at all to float perfectly on the surface has been proven by the many wingless and tailless patterns that already exist. One of the most famous series of wingless parachute patterns has been designed by the Swedish master flytyer Tomas Olsson. On the other hand, when you make larger parachute flies from size 12 and up, then suddenly the wing becomes more essential to keep the much bigger parachute flies floating!

Nowadays, yarn, foam, monofilament, feather flibres, hackle tips and all kind of hairs are probably the most prevalent materials to use for wings on parachute flies. Similar to the thoughts of my very good friend Wayne Luallen, I prefer to divide my wing materials for my dry flies into two categories: 'natural' and 'synthetic'.

Natural hair and feathers are found in a large variety of textures and quality. Feathers from aquatic birds are more popular than those from terrestrial birds because most people believe that feathers from waterfowl shed water very well. You have to realise, however, that a lot of those feathers will mat their barbs together when they get wet (e.g. cdc). Hair may be hard and solid, or much softer in texture and even hollow. Calf tail (curly) or calf body hairs (straight), which are both extremely popular for wings on parachutes, are not my preference. It makes a pretty good wing post and beautiful-looking flies, but the straighter the hair the less compressible and the hackle can slip off easier than with synthetic wing material like poly yarn. I also find it rather bulky when secured on the hook shank which creates too big an underbody for my style of tying parachutes. However, its biggest drawback is the fact that it easily gets waterlogged after a good fight with a nice-sized fish.

Deer hair makes a very good wing on parachutes too but because it is hollow, it easily gets out of shape if you put on too much tension when forming your wing base. Therefore, I prefer to use it for floating emergers instead. In my chapter about the Klinkhåmer Special I go much deeper into the use of my materials for parachute flies.

Gallows tool attached to my J-vice in action

Branko Killer Parachute, one of the best grayling flies in Slovenia

Parachute Sedge made by Staffan Lindström from Norway. He uses melt glue to make this awesome sedge pattern

Another great Parachute fly tied by Branko Gašparin. Branko used quill feather fibres for the wing

One of my Parachute Wildwater Sedges designed and tied at the Glomma river in the 1980s

Parachute emerger that I used a lot in trout ponds in Holland, Germany and Denmark in the 1980s

Klinkhamer Parachute emerger especially designed for stillwater

Beautiful Parachute Danica Mayfly designed and tied by my friend Walter Luzi from Italy

Walter's fly gave me the idea to tie my own Parachute Danica with poly yarn wing

The two Danica parachutes, the left one from Walter, on the right my variant

Beautiful Parachute Parasol Ant tied by Leon Links

Parachute fly and Footprint Dun tied by Oliver Edwards

Upside Down Parachute as tied in the 1980s

Parachute Baetis CDC and foam emerger as used a lot in Denmark by members of the Grayling Society Holland. This one tied by my friend Rob Dings

Same pattern but with tail from Coq de Leon fibres, yellow CDC and orange foam wing

CDC Puff Ball Parachute, tied by Leon Links

CDC Parachute as I like it best

Parachute Ants tied by Leon Links

Parachute fly tied by Roy Christy specially for this chapter

A set of three wonderful Scandinavian parachute flies tied on the Swedish dry fly hook

Chapter 4

Scandinavian Parachute Flies

The Swedish flyfishers in my Scandinavian circle of friends always showed great interest in parachute flies. Some tied really big ones I liked a lot. In the mid 70s Nils Eriksson and Gunnar Johnson began to experiment with hook designs that finally led to the Swedish dry fly hook. It is a well-considered hook, specially designed to imitate large upwinged flies tied parachute style. Gunnar and Nils have written many articles with new ideas and solutions to flytying problems. Gunnar also produced very interesting books with beautiful artwork and was the editor of the magnificent magazine *Flugfiske I Norden* for a very long time.

Both men introduced several radical new patterns. Their knowledge of entomology influenced their design changes to make the hook more effective. They realised that hook point penetration disturbs the water's surface. Contrary to what most people think, their special hook has no holding problems as I can confirm after using it for many years. It also has no hook point penetration of the water's surface because the hook is specially designed to fish it upside down. The point of the hook and the eye are almost in a direct line of pull with the leader. Even when I fished the hook upside down, it did the same trick, exactly like my hook design for my Klinkhåmer Special. Nor were there ever any difficulties in hooking the trout in the hard upper lip of the mouth, nor with hooking large grayling either. However I must confess that I only use the biggest hook sizes, which is exactly what Gunnar suggests.

Finally, after a few seasons of successful testing, they contacted Allan Peterson, a well-known tackle dealer, who was interested and together they contacted Partridge of Redditch who produced the hook in 1979 with the code K3A. Alan Bramley produced the hook in sizes 10 to 18 but I only used it in size 10. Like Gunnar and Nils, I didn't like the small sizes at all. Because the Swedish dry fly hook is discontinued, I couldn't give you the tying steps for the Swedish dry fly. Fortunately, I found a reasonable substitute hook so I can still show you how to tie a beautiful upside down mayfly in the same style which Gunnar and Nils invented. Of course, I still use the old hook, because I have plenty left!

Above left: original K3A Swedish dry fly hook
Above right: Gamakatsu's C16-B Barbless Cripple Emerger fly hook as substitute

Tying a parachute on a Swedish dry fly hook

Dressing

Hook:	Swedish dry fly hook or substitute hook when not available. Size 10
Thread:	Semperfli Classic Waxed 8/0 in tan colour
Wing:	Semperfli poly yarn in white colour
Tail:	Leftovers tips from peccary or Microfibetts
Body:	Fly-Rite poly dubbing in light tan colour (no 19)
Thorax:	Peacock herl
Hackle:	Light blue dun or ginger

Tying steps

The tying can be done in various ways, this is just one method of many.

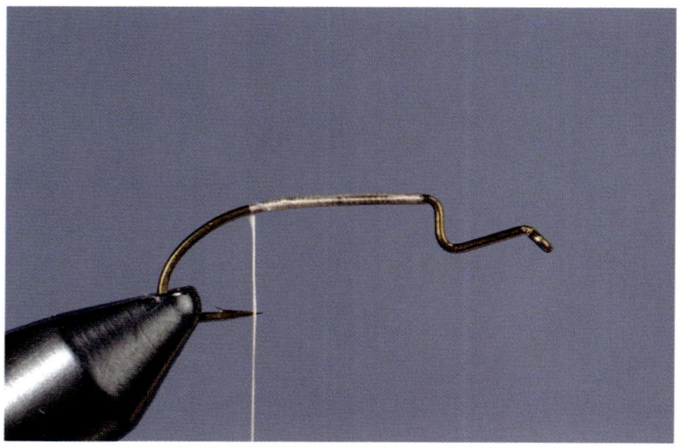

1 Put the hook into the vice and cover the shank with your tying thread.

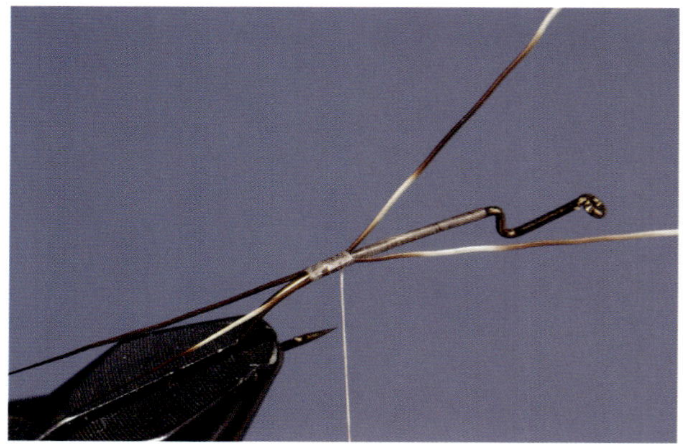

2 Take two nice hairs of peccary and tie them in as a tail. You can also take 4 fibres of Microfibetts and tie two at each side.

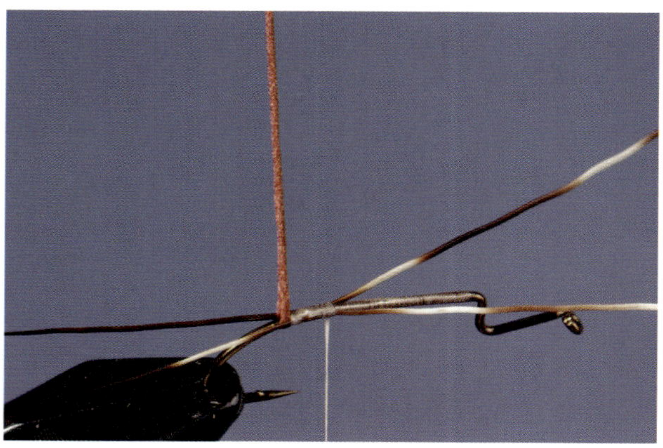

3 Take a thicker piece of tying thread or yarn. Put it in a loop around the hook shank and pull it upwards between the two fibres.

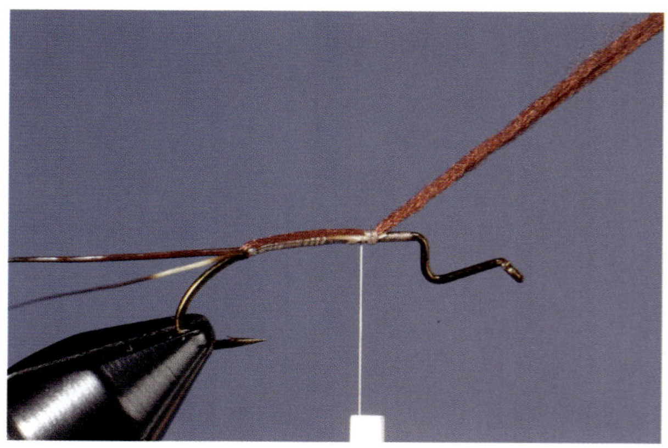

4 Then pull the yarn forward, secure well and cut off excess. Now the tail will stay nicely in a beautiful V-shape.

5 Tie in your hackle as shown above.

6 Apply your dubbing onto the tying thread and start wrapping your body.

7 Build up a nice tapered body.

8 Turn the hook in your vice and tie in your wing as shown.

9 Turn the hook a bit more so the hackling will be little easier.

10 Wrap your hackle to create a parachute.

11 Tie off your hackle with your whip finisher.

12 Tie in your peacock herl for thorax.

13 Wrap a nice thorax with the peacock herl and tie off your fly.

14 Cut your wing into the desired length.

15 Completed fly.

16 Closer look.

Swedish dry fly with two double tails of Microfibetts

Swedish dry fly with three tails each made from double Microfibett

Swedish dry fly in Parachute style with tails made from Microfibett

The Swedish dry fly works well in rapids too

Tomas Olsson

Another famous Swedish flytyer is Tomas Olsson, one of my favourite north European flytyers. We had a lot of the same ideas and thoughts about fly making. He created a whole series of special parachute patterns in the early 1980s. Several of his parachute flies were tied with a completely new technique. I first talked to Tomas when I started to do some research work and discovered that he made some similar looking flies to mine. In the ensuing years, I had great respect for his ideas and tying skills. In the 1990s we met few times at shows in Sweden and I was very happy when Tomas gave me his kind permission to write about his special parachute technique and use his original drawing. The difference between Tomas' patterns, the Swedish dry fly pattern, and my parachute flies are not only in the technique used to secure the parachutes, but also in the length of the wing.

The Olsson melted parachute hackle technique

This unique and very creative parachute hackling technique is developed by Tomas. His technique is the most durable way to secure a 'wingless' parachute on a fly. The method has also been described in the superb book of Darrel Martin *Fly Tying Methods*. The centre of the parachute is melted into a poly yarn post by heat. Tomas told me that his technique produces not only a strong fastening post but it also contributes to the very shape of the parachute. When you melt down the post and press the burned end into the hackle, the hackle fibres will be always pointed horizontally, which results in beautiful-looking parachute flies. Tomas Olsson's technique is unbelievably simple and it is great fun to make such a parachute, I can assure you, because I made quite a few of them! His burning wing technique led me to design several other flies as well. The technique is extremely useful for me when making upside down parachute flies.

Inspired by John Betts

His melted wingless parachute technique was relatively new. Tomas told me that he got the idea after he read an article by John Betts about synthetic material flies in *American Fly Fisherman* magazine. It was a great article, far

Original Olsson fly that Tomas gave me many years ago

Original Olsson fly with Microfibetts tails and a melted parachute at the bottom side of hook shank

ahead of its time and I am truly convinced that John was instrumental in popularising the use of synthetics materials from the late 1970s. I know John as one of the most creative flytyers in the world and am fortunate to have met him at several international shows and fairs. In 1988 John received the Hans de Groot Award at the Dutch Fly Fair for his contributions to international flytying. During the same Fly Fair, we had a great discussion about my LT series of 44 flies mainly made from synthetics. His addiction to yarns, nylon and sparkle gave him the nickname 'Mr Synthetics'. You could find him in all kinds of bridal, lingerie and craft stores. He introduced me to synthetic organza and Microfibetts, both of which I use a lot in my style of flytying. The basic material for the Microfibetts are the tapered nylon fibres of white sable artists' brushes. It's one of the best synthetic tailing materials I know today.

In this article John had described the melting property of poly yarn, and how he made extended body flies by melting the end of a piece of poly yarn. All he needed was a lighter and yarn. Inspired by this article, Tomas adapted the idea to create his melted hackling technique. You will quickly discover that the same trick is very useful for several other patterns too. Tomas wrote to me that he designs and uses these wingless parachutes primarily to imitate emerging patterns in the Scandinavian rivers and lakes. The success of the patterns is enormous, provided your long-distance eyesight is perfect. Tomas' own account of the pattern development explained the drawback with previous methods for parachute hackles, which was the lack of firmness obtained when winding the hackle onto the fly. Most parachute techniques call for wrapping the hackle around the base of the wings, and in the case of the hackle tips wings, he found the stems were not strong enough to provide a firm and durable hackle base. With this in mind Tomas tried to find a better method and he did.

To explain this excellent method, I pass on the tying technique as Tomas sent it to me, together with his original dressing and lovely drawings.

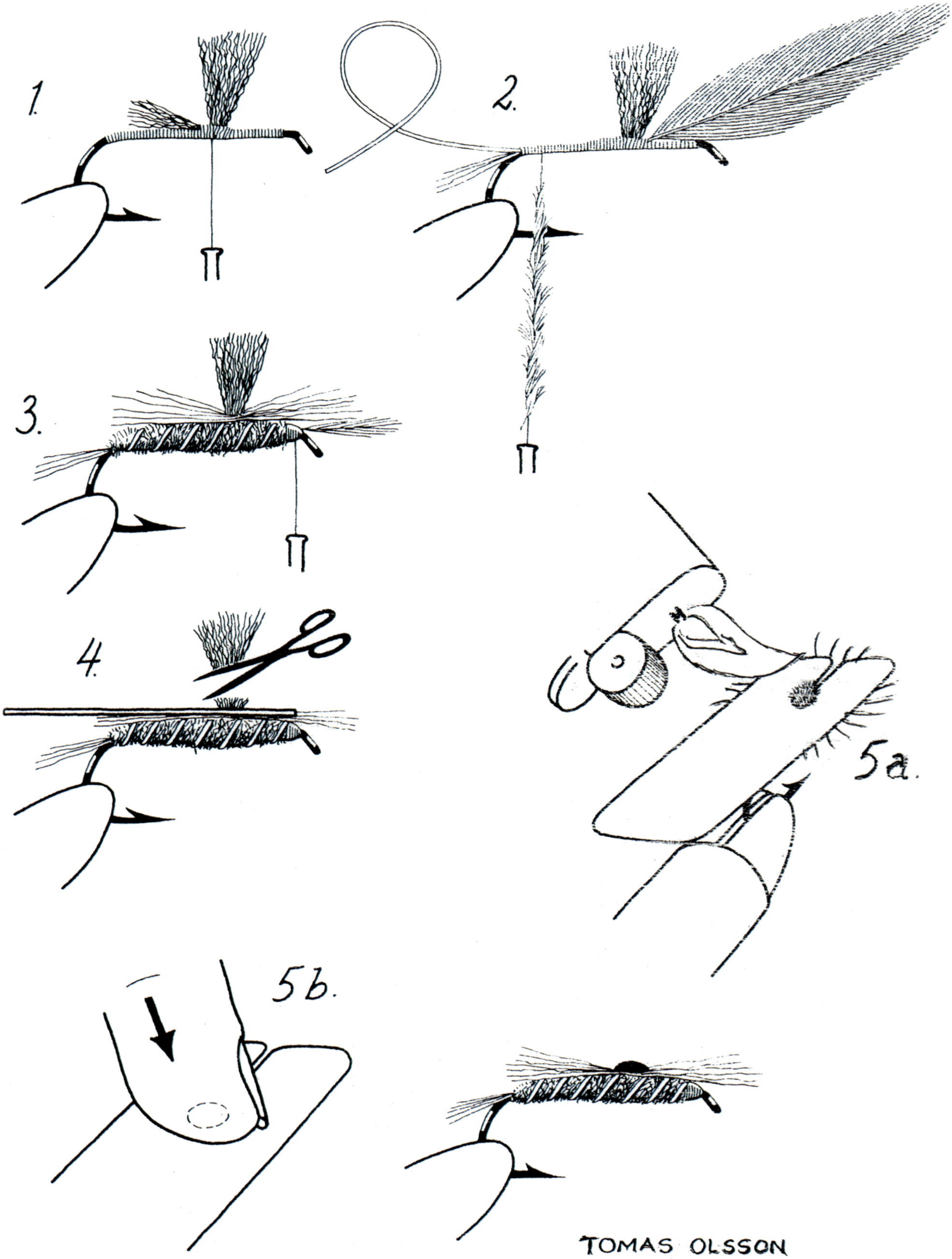
1.
2.
3.
4.
5a.
5b.
TOMAS OLSSON

Tying a classic Tomas Olsson emerger using his original drawings

Hare's Ear Parachute Emerger Dressing

Hook: Dry Fly hook 10-18 fine wire. (I suggest Daiichi 1180-1182)
Thread: Semperfli Classic Waxed 8/0 in tan colour
Wing: Semperfli poly yarn tan or grey
Tail: Ginger hackle fibres or Microfibetts
Ribbing: Gold tinsel fine
Body: Hare's ear dubbing
Hackle: Ginger dry fly hackle
Special equipment needed: A metal or cardboard hackle guard to protect the hackles from burning

Tying steps

1 Put hook in vice and cover the hook shank with your tying thread. Cut off a 3cm piece of Poly yarn and tie it in with wing up.

2 Cover the poly yarn with your tying thread. Tie in the tail, ribbing and hackle. Then apply your dubbing to your thread.

3 Wrap the dubbing forward and secure it with your ribbing. Create a wing post with your tying thread. Then wrap the hackle around the wing post and secure it just behind the eye of the hook and tie off with your fingers or whip finisher.

4 Put the yarn in the centre of your melting tool and press the guard down so it protects the parachute hackle from burning. Then cut off the yarn about 2mm above the hackle guard.

5a Melt the little poly yarn stump with a lighter.

5b Wet your finger and immediately press the molten stump down with your forefinger to form a flat bead on the hackle.

6 Completed fly.

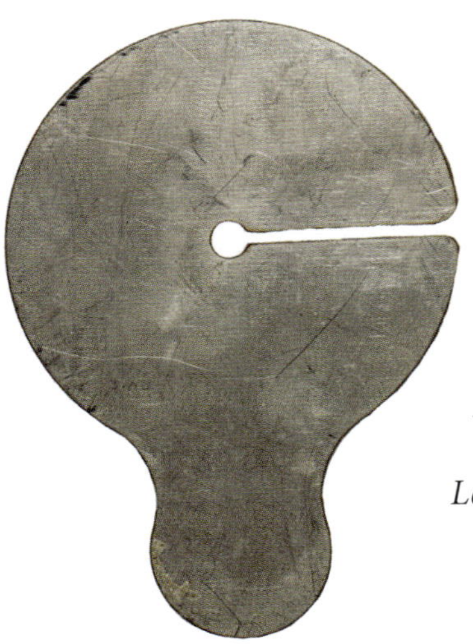

Homemade hackle guards for burning the wing. Left a metal one, right just made from cardboard

How to burn a synthetic wing using the famous Olsson technique

1 Grab any parachute fly with a synthetic wing from your fly box. I recommend poly yarn wings.

2 I chose a big fly so I place the large hackle guard over the wing to protect the hackles from burning.

3 First cut the wing so the burning process will get a lot easier.

4 Result after cutting.

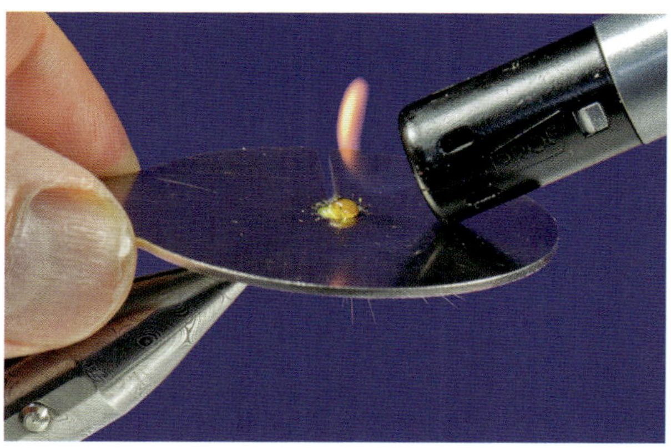

5 Melt the little poly yarn stump with a lighter.

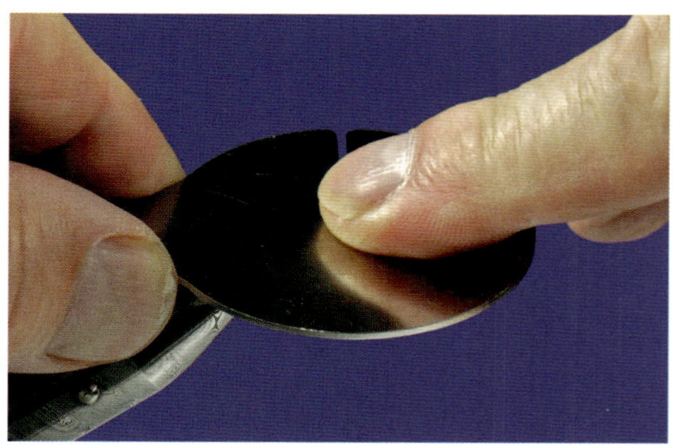

6 Wet your finger and press the molten stump down with your forefinger into the hackle while it is still hot.

7 Result after pressing: a nice bead is all that's left from your synthetic wing.

8 Because I can barely see parachute flies without a wing, I put a little drop of Tulip (Glow in the Dark) fabric paint on the small synthetic bead and let it dry for a while.

9 Final result after drying. A perfect waterproof fluorescent dot that is extremely durable and it makes your wingless parachute fly clearly visible.

Tomas' pattern, made in my way with Tulip painted sight indicator on top

Nice grilse in Newfoundland took a Tomas Olsson variant

Universal Baetis tied by Leon and me

Chapter 5

Universal Baetis

Over the years the parachute fly has gained in popularity, primarily due to its attractiveness to fish, but not solely. Articles and books gave it publicity and in the 1970s there was a lot of discussion about the various types of flies, including dry flies. There was growing recognition about the importance of the respective parts of the flies, the body, the outline, size, wings, hackles and even about the colours of flies. The high catch rates certainly brought parachute flies to attention.

Important books

In *A Modern Dry-Fly Code* (1970), Vincent Marinaro stresses the importance of a clearly visible wing. His observations convinced him that winged flies were superior to wingless flies and that wings drew the attention of fish, more than a fly's body. Several authors didn't agree with him and pointed out the significance of the body; others said that the legs left an important dimpled pattern on the water surface.

Because I totally went my own way in my flyfishing and even more so in my flytying, I developed my own thoughts and beliefs. That is why I completely understand why some tyers wrote so differently about certain patterns and techniques. I liked the way some authors, flyfishers and flytyers were clearly doing things differently to others and when they responded to their critics, the more I could believe in their arguments and explanations. Maybe their passion for their ideas really attracted me and pulled me further into their narratives.

I have so much respect for all those flyfishers who didn't just go into the straitjacket of the famous ones.

Doug Swisher and Carl Richards examined fascinating subjects in their influential book *Selective Trout* (1971). After thorough research, their conclusion was that many aspects of an artificial fly are important, especially when fishing for selective fish. About dry flies they say that fishermen should take notice of 'characteristics of size, shape and colour, but also such peculiarities as how the fly floats on the water and how it behaves in its environment.' (*Selective Trout* p.24) So, a trout must be led to believe that the presented imitation is a real insect. When a freshly hatched insect approaches the 'window', the first element

Original book cover from 1971

a fish can distinguish is the wings of the fly. In the view of Swisher and Richards, the wings act as an important 'trigger' for the fish to start its rise. The fish, however, only makes its final decision to take when it approaches the artificial and finds that nothing is 'fishy' about size, wings, colour, body and proportions.

Swisher and Richards recommend various flies for selective trout, especially their own 'No Hackle flies', but they are also positive about parachute flies. Until the 1970s, people used parachute flies in larger sizes only. Swisher and Richards fished them successfully in small and tiny sizes too and they even tied them down to size 28 hooks.

They are also clear about the objective in using a parachute hackle. It is solely meant for delicate presentation and to obtain the correct position on the surface, and not – and they emphasise this – for improving flotation. The parachute hackle, therefore, should be really sparse, especially when you tie slim flies. Otherwise it would spoil the silhouette of the fly, as the fish would not be able to see the body and the wing.

Their ideas about the use of poly fibre in flytying was close to my heart. I love their book in many ways and I think every flyfisher and flytyer should read it. There is a beautiful revised and expanded reprint of their first edition published in 2001.

Lessons learned through addiction

I love Denmark because I am always sure of having an enjoyable trip and that's the main reason I still visit it twice a year since the early 1980s. In the past when I lived in Germany, because of my army duties, I visited the country 5-6 times a year. Now I am retired, I normally make longer trips of 3-4 weeks during each spring and autumn. Each trip to Denmark gives me new inspiration and ideas regarding flyfishing and flytying.

Although I am quite familiar with the huge rivers in northern Scandinavia, I also like to fish small brooks and streams and Denmark is an excellent place for it. I think my pleasure has a lot to do with my past when as a kid I fished quite a few small and similar-looking rivers in the south of Holland. There are countless little streams in Denmark and most of them still have their natural watercourse. It is wonderful to see how they meander quietly through the varied landscape that is so characteristic of Jutland. Gently sloping, here and there groups of trees, fields, meadows and wetlands. Beautiful farmhouses in the middle of natural shades in all kinds of nuances.

Those deep eroded meanders produce perfect lies, shelter and hiding places for fish. The current sometimes runs quite fast and has many marvellous feeding seams between the beautiful weed beds and high sandy banks. The conditions are perfect for insects and caddis and mayflies still abound. At the end of May, you will find large populations of mayflies which attract not only fish but also large groups of dry fly fanatics. In the spring and autumn many birds, on their way north or south, take rest in the fields and bring extra joy to the fishermen. Birds of prey are hunting day and night. Later in the season when terrestrials fall into the water from the overgrown banks, you sometimes see those pools boiling with activity. I can watch it for hours in search for a new trophy fish. When the weather conditions are perfect, up to ten large fish a day are possibilities. Of course, I can go for numbers and concentrate on the smaller fish but it is not my intention to do so, especially not with my pretty large artificial flies that I am using. It can damage small fish too easily.

A wonderful place to be

In the 1990s I started with my classes for the Danish FFF which they organised at the Gudenå Camping Brædstrup at the River Guden in central Denmark. In those days, the camping was owned and operated by Anna and Henning Gregersen and they were the best hosts imaginable for an event of this size. The location was chosen because of the great accommodation, including a nice dining room and a large and separate presentation room for our workshops, classes and lectures, so we could leave all our flytying materials on the tables while we were fishing. We stayed in simple but modern cabins and made further use of all the camping facilities, including free fishing on a stretch of a few kilometres of the wonderful River Guden. Anna and Henning always served an excellent breakfast and superb dinner and allowed us to prepare our lunches ourselves. In the evenings, there was lots of flytying, a few drinks, incredible tales and lots and lots of laughter. The event went so well for several years in a row that I started organising one myself for the members of the Grayling Society Holland for quite some years. After nearly 10 years my secretary, the late Ben Bod, took over and also organised the trips for our Grayling Society in an unbeatable way.

Beautiful meandering little stream in Jutland

I remember very well that when we arrived, the most eager fishermen in our group naturally tried to catch a fish as soon as possible. Nice trout and a few graylings were tempted by dry flies and nymphs.

A really beautiful moment

When I asked Leon Links to compile a top 5 of the most beautiful flyfishing moments of 1996, he told me that it would certainly contain the following memory of this particular day: *'Hans and I take a rest on a rickety footbridge over the Gels Å. We light up a smoke and look around. Fish regularly rise between the white flower-covered, swaying weed beds in the calm stream. Around us fields, in the distance a farm, groups of trees. Not far from us the knocking of a woodpecker begins and everywhere, really everywhere, mayflies flutter around us and above the water. A large specimen floats under our bridge. Rising fish just don't notice him. We grin at each other. Beautiful face. Just visible to us, it disappears unexpectedly in a gully, some thirty meters further on. Gone... What remains is a circle that is rapidly expanding'.* I remember that same moment very well and although we were not even fishing in that particular moment, it really made my day and it is all a part of flyfishing for me too.

The rise of the Universal Baetis

The Universal Baetis (UB) is a fly that emerged during a trip of our Dutch branch of the Grayling Society to Denmark in 1997. It was a spring trip and while fishing we were frequently confronted with hatches of Baetis *Scambus* (small dark olive dun) and other Baetis species like Baetis *Rhodani* (large dark olive). It was raining but the Baetis kept hatching. We agreed that we lacked an adequate imitation, which made the catches rather disappointing. We mainly fished for grayling and we noticed that the fish were feeding very selectively.

In the small brooks of Denmark you can sometimes catch quite big fish

The Universal Baetis with quite a long wing

All flies of the wrong size, wrong shape, wrong colour or tied in different proportions were steadfastly refused. So, in the evening after the main tying sessions, Leon and I started figuring out a fly that would work. Our intention was to create the best possible pattern imitating the characteristics of the insects that were hatching during the day. Of course, we had caught some of hatches and larvae to investigate during our tying sessions. To determine the correct colours for the dubbing we made grateful use of the colour cards in the Borger Color System booklet which helped us select from our extra fine poly dubbing collection from Fly-Rite.

This resulted in a fly that worked much better than the flies we had used previously. We thought we had found the perfect formula for a slender parachute fly that simulated the delicate naturals really well and we hoped that the new fly that we designed together would be taken without any hesitation the next day. To start making our fly we took a standard dry fly hook and tied in a small gold yellow tag as a special trigger. With the tiny tag, it was also a lot easier to give the fly these essential stiff and relatively long split tails. By using vintage orange pure silk as ribbing we both believed that we had found the most perfect colour to create the body segments to match the real insects. The abdomen and thorax we had both selected using the BCS system and it quickly led us to the colour 42 in the Fly-Rite colour range. Please realise that we selected the colours by thinking fully wet and even checked this by dropped our dubbing choice in a bowl of water first. The hackle for our parachute was not that difficult to choose and we kept it flexible by picking a few different colours of dun. We experimented when fishing with a hackle in light, medium and dark dun colour. For the wing, we took a nice shade of light grey poly yarn.

The fly did very well but after a small modification it worked even better, namely by extending the wing. During a walk along the river, Leon and I noticed that the closed wings of the insects drifting on the water cast a clearly visible reflection or shadow on the water, which we believe could be a key trigger. This idea was confirmed time and again while fishing the narrow fish-rich Danish rivers in those days. We noticed that the effect worked best on days with changeable weather and especially during sunny or rainy moments.

Over the years, we also discovered that the height of the wing is very important to the overall appearance of the parachute fly. We strongly believe that people often tie wings much too short in general anyway, probably because they are influenced by the size of the hackle or hackle fibres.

Let me give you a good example: when Leon and I developed our successful 'Universal Baetis Dun' prototype, we mainly succeeded because we tied the wing much longer than we usually would do. We both felt that the wing was much too long when our new fly was completed. Leon had already taken up his scissors when I suddenly remembered his 'Fragile Darter or Crazy Sedge'. I told him to stop and that we needed to give it a try with the longer wing first. When we did so the next day, and saw how our newly developed artificial fly appeared while floating among the natural insects on the surface, we both were amazed that the height of our wing was actually no longer than the wings of the real insects. A clever strategy that you may wish to employ, therefore, is to tie your wings a little longer next time. You can always cut them while fishing.

Afterwards Leon and I continued to use this pattern during Baetis hatches and always with very good results. We had the definite impression that the longer wing was important and that its reflection worked as an extremely effective trigger. In that and the following years we found the pattern successful on many other rivers as well. That's why we baptised our mutual creation the 'Universal Baetis'. The development of our new fly also forced us to use the word 'silhouette' instead of shape and we still use that term today.

Two beautiful Universal Baetis tied by me and Leon (note the extra long wing on the right fly)

Bunch of Universal Baetis tied by Leon and me

It might create many a serious problem to fish between high grasses, nettles and thistles, but I love it

The deep corners of a small Danish brook can produce very big fish

Universal Baetis in all its glory. Notice the long wing!

Tying our Universal Baetis (as we designed it in Denmark)

Dressing

Hook:	Daiichi 1180 size #14, #16
Thread:	Semperfli Classic Waxed thread yellow, olive or dark grey 18/0
Tag/butt:	Fly-Rite dubbing colour #17 Golden Brown
Tails:	Four dark grey Microfibetts split 2 by 2
Ribbing:	Pearsall's Gossamer 6a or Semperfli Pure Silk Vintage Orange
Abdomen and thorax:	Fly-Rite dubbing #42 Dark Olive Brown
Hackle:	Dark or medium dun cock hackle
Wing or post:	Light grey poly yarn

Tying steps

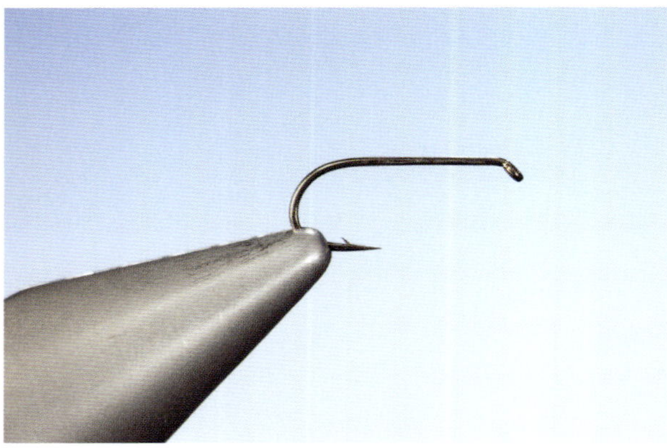

1 Pinch down the barb of the hook if you wish and insert the hook carefully into the jaws of your vice.

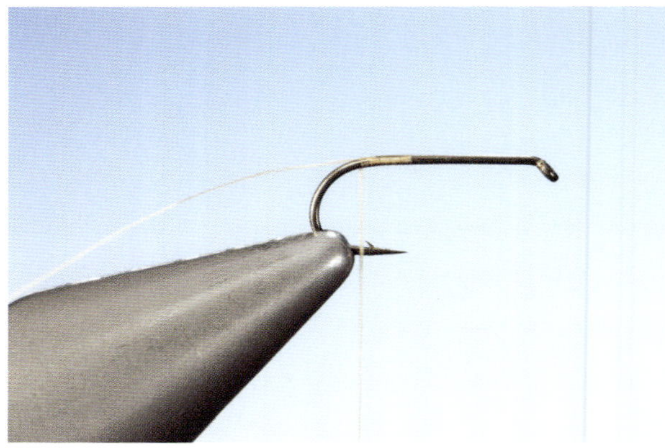

2 Put on some yellow tying thread just above the barb.

3 Wax the thread.

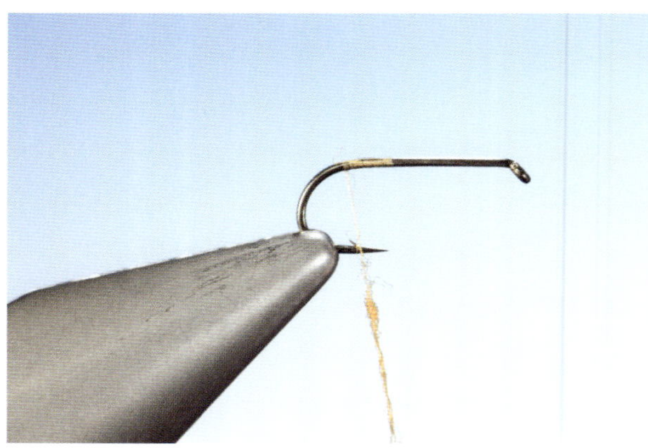

4 Add just a little bit of the #17 Fly-Rite to the thread.

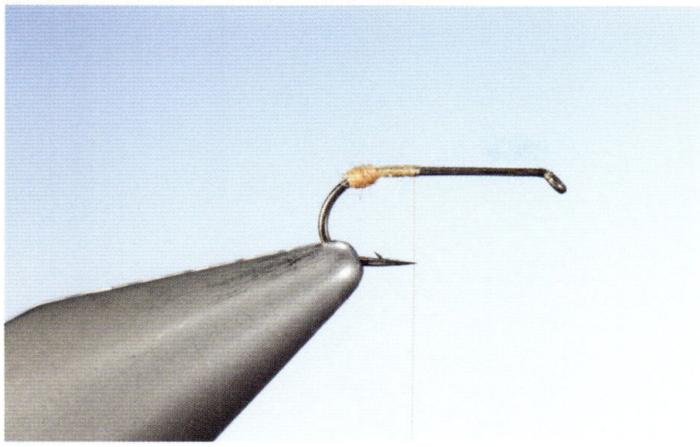

5 Create a little butt of dubbing just in front of the bend of the hook and wind the tying thread slightly forwards towards the hook eye.

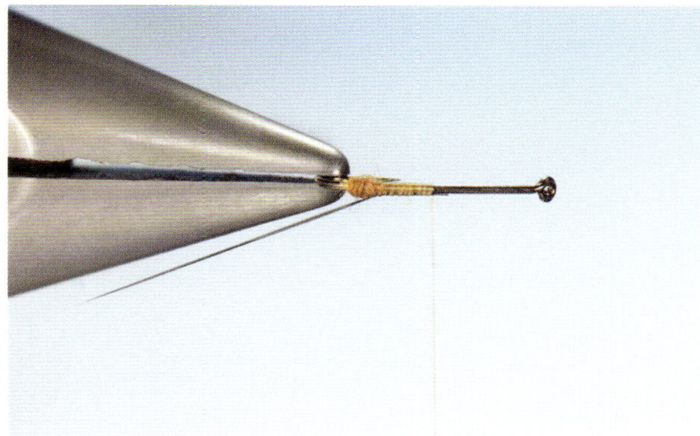

6 Tie in two Microfibetts at the nearest side to you and secure them gently while adjusting the length at the same time. Carefully secure the tails with your thread until you reach the little dubbing butt. Clip off the waste ends of the fibetts.

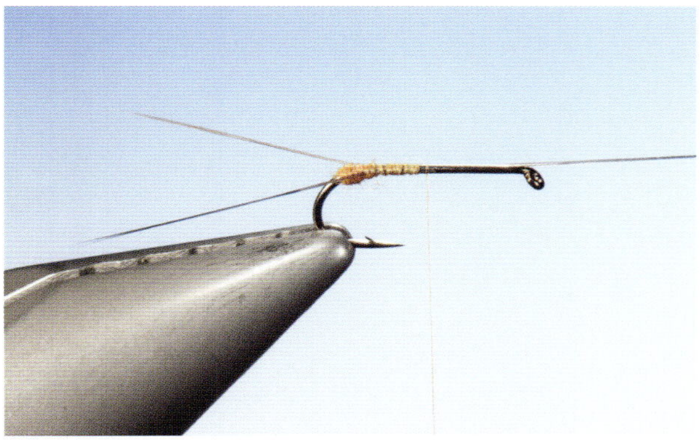

7 Repeat with two Microfibetts on the far side of the hook shank. Secure the position of the tails by firmly wrapping your tying thread a few extra times against the little butt. Wind the tying thread slightly forwards. Clip off the waste ends.

8 Tie in the poly yarn and secure it at ¼ of the shank length by figure of eight wraps. Wind a thread base around the poly yarn post. Tie in the orange silk for ribbing just in front of the butt or tag.

9 Tie it in the hackle in front of the post, dull side facing the shank. Carefully form a post base on the poly yarn by winding your thread around both the hackle feather and poly yarn together. Snip off the stem (or quill) of the hackle.

10 Wax the thread and add dubbing to it, but sparsely.

11 Dub a nice tapered body starting from the butt and moving along the hook shank until just before the poly yarn post.

12 Wrap the silk ribbing in the opposite direction. Secure the silk in front of the post with some extra windings.

13 Apply wax and a little bit of dubbing to your thread and wind a nice thorax. Then bring your thread around the post and let it hang on the near side of the fly.

14 Wrap the hackle around the base of the poly yarn post. Begin *at the top* of the thread base, and make each successive wrap down the base and ensure that each wrap is made *under* the previous. Secure the hackle with three or four wraps of thread. Whip finish.

15 Completed fly.

The Universal Baetis, as deadly as can be

Universal Baetis in sizes 12-14-16 and 18, all with long wings

Original Klinkhåmer with a beautiful mixed tone body from different colours of Fly-Rite dubbing

Chapter 6

Klinkhåmer Special

Forty years of dry fly success

When I re-read my earliest flyfishing diaries, I enjoy the old stories; they keep my memories fresh. I find that as I get older, I tend to read them more often. Sometimes I just read the fishing parts and get a lot of pleasure from the notes about my trips, but there are times that I only search for the comments I wrote about my flytying experiences. It is very important to know why and how I came up with a special pattern, and why I chose certain materials or changed dressings and proportions, sometimes dramatically.

Since the early 1980s I have designed and improved many patterns and it is very handy to have good sources of correct information about those flies. I also like reading similar stories from fellow anglers, or a few of my closest friends and we often swap our diaries and notes. Many flies have been designed as 'fancies', others arise accidentally, and a lot are close copies of insects, but I often miss the stories behind the patterns. I like to know the thinking of the originators, and how they discovered certain tying tricks or techniques. Sadly, many flies lose their names and change so much that their histories and original dressings just fade away after a few years. My biggest strength in analysing flies is the fact that I can take a detached look at flies because in my home country of Holland there is not really a flyfishing history nor a history about fly patterns at all.

Setting the record straight

It is no secret that parachute flies have been my favourite dry fly patterns for more than forty years. It was the introduction to the 'Rackelhanen' that set me free from all traditions. Not only did it make me become innovative, it allowed me to think outside the box and forced me to use a much wider variety of materials than I had ever tried before. I like it when flytyers are inventive, flexible and creative, but this doesn't mean we should change the true stories behind fly patterns. I enjoy telling people where I got my inspiration from and I always acknowledge those who gave me an idea that led to one of my new patterns. In this age of word processing, original text can be easily altered or changed. I continue to experience this with my own patterns; sometimes to the point that I am not able to recognise my original fly anymore.

It has happened more than once with the Klinkhåmer Special. And on the other hand, there are a lot of exact copies of my Klinkhåmer fly but with completely different names. I have seen the Klinkhåmer Special tied on straight-shank hooks, with tails, with different wings, and with abdomens thicker than the thoraxes. Such changes create only misunderstandings and for many years I have received countless letters asking questions about the Klinkhåmer.

I think this chapter of my book is an excellent opportunity to correct a lot of misunderstanding about the Klinkhåmer and to share with you a few previously untold secrets as well. Therefore, here is the definitive update about the best dry fly pattern that I ever invented, and hopefully we can celebrate together four decades of dry fly success with my Klinkhåmer Special.

Past, present and future

My first Klinkhåmer Special landed in the surface film of Norway's mighty Glomma River on 27 June 1984, and a lot has happened since that particular day. Since my first publications in 1986 I have read a lot stories about the Klinkhåmer written by others. Some articles were really good, but most contained misunderstandings about my thoughts behind the fly. That is sad; even when those authors gave me a lot of credit for the pattern, there were always a lot of readers who didn't learn how to properly tie and fish the Klinkhåmer Special. It hurts me when I see a good pattern described incorrectly. Some stories about the Klinkhåmer described clever improvements and awesome colour variations, but unless they affect the fly's durability, floatability or create a better silhouette, I don't see them as real improvements.

With so many variations of my original Klinkhåmer out there, it is no wonder I get phone calls, e-mails, and questions about how to tie a 'real' Klinkhåmer Special properly. The good thing is that there were many flytyers who were experimenting with my fly and in very exciting ways. By the time you reach the end of this chapter, I hope to have cleared up the obsure areas where most people make their common mistakes when tying and fishing my Klinkhåmer Special. I also will inform you about my latest change in my dressing, and why I did it, which is a serious update on the Klinkhåmer Special version 2.0 or even 3.0.

Sharing the credit

Before you give me all the credit for the Klinkhåmer Special, you should know that I wasn't the only one who struck upon the idea of designing deep surface hanging, parachute-style flies. Unknown to me at that time, Tomas Olsson, a famous Swedish flytyer, had created a similar pattern just one year earlier. With his special melted technique, he developed a similar but wingless fly. About the same time, in 1986 in the United States, Roy Richardson developed an identical emerger but with a tail and two wing cases over the thorax without knowing about our flies. Mike Monroe, also from the United States, tied a similar fly four years before any of our patterns existed; all this at a time we in Europe hardly knew what was happening in the American flytying scene. Mike used a normal

An early Klinkhåmer totally destroyed by salmon

My first Klinkhåmer tied with calf tail as wing

The stomach contents of this grayling were responsible for the design of the Klinkhåmer Special (original 1984 photo)

The birthplace of the Klinkhåmer Special: Kvennan Camping as it looked in 1984

Typical Norwegian grayling. We caught so many in Hedmark between 1984 and 1996 (© Hein van Aar)

Beautiful Arctic grayling caught in Yukon Territory, Canada 2003

Another Arctic grayling seduced with the Klinkhåmer Special in Yukon, Canada

hook and called his fly the 'Paratilt' and published an article about his special design in the July-August 1979 issue of *Fly Fisherman* magazine. Unless new information is uncovered, I think all these other tyers deserve as much credit for their creations as I have had for the Klinkhåmer Special.

Naming of the Klinkhåmer Special

Never have I seen a pattern that has been misspelled as often as the Klinkhåmer Special. At first I could understand this because in the mid 1980s, all my writings were done in long hand. Not possessing the best of handwriting skills, I could easily see how a misspelling could occur here and there. However, I began to notice that in the 1990s, several people were changing the name deliberately. To this day, it is beyond me why this has occurred!

In Germany, they call it the Nordischer Hammer or Klinki. In the United States they seem to prefer the Clinck or Klink, and I often get questions about all kinds of 'Hammers', most of which I have never heard of before. I have seen Pinkhammers, Yellowhammers and even Bluehammers to name but a few, and all were named by others. However, over the years I finally started to accept the nickname for myself of Klink, because so many people just call me Mr Klink. Of course, I can't deny that I swell with pride when the Klinkhåmer Special gets so many great reviews but I also feel a bit sad when I read so many mistakes. For the record, I remain most proud of this fact: it was none other than the late Hans de Groot who coined the name that actually gave the fly its moniker. Over the last 40 years I have received approximately 300-400 letters asking about how I came to choose the name and, more importantly, what its meaning is!

There is a large mountain in Sweden called the Hammer, very close to where I used to fish a lot in my early days and several people spread the rumour that that's where the name comes from. Other people say the fly hits so hard, it's like a sledgehammer. None of this has anything to do with the real meaning behind the name! Fake news! The original name was the LT Caddis, which was simply number six from my large **LT series*** developed in Scandinavia between 1981-1991. Hans de Groot, Franklin Moquette and Ton Lindhout came up with the striking name of Klinkhåmer Special while editing my story for the *Nederlandse Vliegvisser* and it has stuck ever since. We were all members of the editorial staff of our national flyfishing magazine at that time. They didn't think the name LT Caddis was appropriate for such a powerful fly, so they took the liberty of renaming it – and at the same time maybe teasing me a bit. I was absent from that particular editorial meeting and they later confessed to me that they wanted to give the name a bit of Scandinavian flair. Fishing in Scandinavia was my first preference and I was addicted to fishing there. Hence, the name was born! I strongly believe that the new name was crucial to the fly's popularity because in every country I have been everybody is able to pronounce it properly. Even Chinese and Japanese people pronounce it correctly.

*The LT series stands for the Light Tan colour number 19 from Fly-Rite poly dubbing because all the bodies of the 44 patterns in my LT series were made with the same colour of poly dubbing. The LT Caddis was number 6 in the series but tied without a thorax. By adding a peacock thorax, the fly was suddenly looking different and with the brand-new name I added it as number 7, the Klinkhåmer Special, in my LT series.

Real thoughts behind the Klinkhåmer Special

Personally I believe that most mistakes that are made while tying Klinkhåmers occur because people are stubborn, don't read the instructions properly, and tie my fly from looking at pictures and don't listen to the thoughts behind the pattern, and why I use certain materials. I have dozens of examples why Klinkhåmer Special

Our holiday house with veranda over the Glomma river which we used for nearly 15 years

For this size grayling, I keep returning to Scandinavia year after year

River Glomma with green circle marking the birthplace of the Klinkhåmer Special in 1984 (© Hein van Aar)

Beautiful coloured Scandinavia grayling caught with a 2023 version Klinkhåmer

flies fail and most I can refer back to people using different materials to the ones I suggest in my dressings. I think before attempting to tie a Klinkhåmer Special properly, you would benefit from knowing what I was thinking when I created the pattern. Quite a bit already has been described and explained in my Parachute chapter. My great preference for parachute flies was definitely the start of the Klinkhåmer thinking progress just a few years later. It was the foundation of the development to keep my crazy iceberg shape fly floating so extremely well.

I repeat that when fishing for grayling with dry flies, I believe that it is a combination of the shape of the mouth, the speed of the rise, and their way of taking the fly that causes a grayling to push away a regular floating pattern. A parachute-style dry fly reduces this problem, and an emerger that hangs deep in the surface works even better. I proved this theory many times after designing the Klinkhåmer Special. The iceberg shape, or deep surface-hanging emerger style in which I tie my flies, solved these problems and improved the catch rate incredibly.

Even flytying expert and master flytyer Oliver Edwards told me that his catch results went up a lot after he started using the Klinkhåmer or Klinkhåmer style flies. I also observed that at least nine out of ten grayling were hooked well in their upper lip when using Klinkhåmer flies.

Flyfishing with deep subsurface hanging parachute flies offers many benefits. A well-tied parachute fly lands perfectly on the surface and floats wing upright the entire drift. Because of the 'iceberg shape' in which most of the mass of the Klinkhåmer Special is below the surface, the upper part of the fly is vertically stable. If tied correctly and with the right materials, this fly floats through riffles and strong currents without tipping on its side or sinking.

For me, the idea of a parachute fly is to create a strong fly silhouette for a fish. I think that it is this deep in the surface, hanging presentation of my Klinkhåmer Special that allow the fly to imitate all types of emerging mayflies, caddis and even some terrestrials. The appearance of an insect caught in or below the surface film makes a tempting target for trout and grayling, but the

Ina trying a Klinkhåmer Special at Sterkfontein Dam in South Africa

Barbel in Spain which couldn't resist a black Klinkhåmer Special

position and shape of the hook also allows you to hook fish more frequently and securely. Even Atlantic salmon, notorious for merely bumping flies with their noses, are at certain times hooked more frequently with the Klinkhåmer than with traditional bomber flies or bugs. Seatrout sometimes can go mad for them too but they prefer slender ones.

When we tie imitations of insects, only a few points seem important. We consider the size, shape, colour, and mobility of the naturals, and replicate these elements in our patterns. But I think there are more points of interest, and I search for them every day. For example, several years ago I started thinking differently about fly design, and today I teach my flytying students to consider the silhouette a fly produces. I don't mention the shape anymore. Thinking like this suddenly makes the length of the wing extremely important.

I try to look and see things in a different way. Wind can improve mobility and we already include that element in our flytying. But how many tyers take into account the effect of the sun on the fly as it rests in the water? I now realise more and more the importance of silhouette and the great influence of the sun and rain on the colours from patterns like spiders and other flies that you fish just underneath the water film. I can offer no other explanation as to why some selective fish will only take a certain fly when the sun sets or rises or just disappears behind a dark cloud. After fishing Klinkhåmers for forty years I changed my mind radically about the length of the hackles within the parachute too, and that is why I have also a good number of Klinkhåmers with extremely long hackles in my fly boxes today.

The original materials

We are now approaching the main reason for writing this chapter. I have seen too many mistakes in other articles about the Klinkhåmer Special, and I want to put it all right. Not everyone will accept my explanations, but I will try to explain my personal reasons for using specific materials, and what will be the best substitutes. I will tell you why I designed my own curved, wide-gape hook, and why I developed a completely different tying technique to secure the parachute hackle. There have been many changes in the flytying world over the past forty years regarding my fly, and I still use almost the same materials today as I did when I first developed the fly. I have seen many variations, and have tried to make dozens of improvements myself, but only a few proved better or more effective than the original. I have actually three original types of Klinkhåmer flies, which I will describe in this chapter.

The hackle

There is a huge difference for me between tying small or medium-sized Klinkhåmers and huge ones. Despite what some people say, you need the best hackle quality that you can find when you tie small, slender or normal size Klinkhåmers or any other parachute flies. A 'light', normal or small dressed parachute fly (size 14 down to size 20) floats in the surface because of a stiff, balanced, and well-tied horizontal hackle. A size 12 hook is a borderline case. A tail can be helpful with a normal parachute fly with a straight hook but isn't really necessary. However, for my big size Klinkhåmers (size 10 up to size 1) it doesn't really matter if you use cape leftovers or even low-grade hackle quality. The big sizes all float on the poly yarn wing and the hackle hardly has anything to do with its floatability. Several people I know claim the best hackle for a parachute fly would be from a good quality saddle cape but I totally disagree (except when using Coq the Leon saddles) because top saddles usually have very short fibres and I don't like them short on any of my parachutes.

For me a good quality parachute hackle must have a strong, flexible, but not too thick quill. The fibres should not be too short and must be as stiff as possible. The hackle in general should not be too short either, allowing you to make a good number of extra wraps when needed. I can't give you any exact sizes or numbers of wraps; I know instinctively through experience and practice over time. Wrapping a parachute hackle is a completely intuitive operation for me. If I give detailed measurements, people will rely on that and will take it as standard. I can only tell you honestly that I tie my parachute flies 100% by instinct and with very varied wing and hackle sizes. Every large parachute fly tied on exactly the same hook size leaves my vice with a different size of parachute hackle and wing. Sometimes the thorax differs in size or thickness too. So I do not keep myself strictly to proportions when tying my medium or large Klinkhåmers.

I only have some dislikes: for example, I don't like it when the wing is too short or too close to the eye of the hook, or when the hackle fibres are too short, or when the body is too thick, uneven and not tapered. I only tie my parachutes equal from sizes 14 down to 20, simply because these sizes are a lot closer to the naturals and they tend to match the hatch. Small parachutes flies and Klinkhåmers are great to use when fish are very selective and hatches are on the go, while big ones I mostly use as a fish finder, to seduce the monster fish to do silly things and to prevent hooking too many small fish. A dish of water on my tying desk was and still is indispensable for me to find the correct hackle quality and fibre sizes.

Although I was fussy about the size of hackle fibres in the mid 1980s, I soon learned that oversized hackle fibres on my Klinkhåmers and other parachute flies wasn't a problem at all. The best proof were my catches and I often caught my biggest fish when I used a fly with pretty long hackle fibres. Many people must think that my Klinkhåmers are terrible flies with such huge oversized hackles, but I can assure you that the fish doesn't care how large the fibres are. In short, it is hard to over-hackle the fibre length in a Klinkhåmer Special: all you need to do is to believe in it like I do. Sadly a lot of people tie Klinkhåmers much too small, and small Klinkhåmer flies catch a lot of fish, but you catch many small fish too. With bigger flies, you prevent that problem.

In South Africa, they have a tiny dry fly called the Para-RAB (Red Arsed Bastard or politely renamed Rough And Buoyant) developed by Philip Meyer in the 1990s. It is an iconic South African dry fly with probably the most over-dressed parachute form I have ever seen in my entire life. The special halo effect created by squirrel tail hairs is the main trigger in this fly. There are lots of different versions of the Para-RAB today, but they all have one thing in common, a large halo 'hackle' and a slim body. I fell in love with the pattern at first sight and I can assure you it works amazingly well and not just in South Africa.

When I tie my own fishing flies in a group with other tyers, and we put our Klinkhåmers together, mine always seem larger because of the hackle and longer or thicker wing sizes, even when we tie on the same hook sizes. None of my Klinkhåmers is the same and this is a deliberate part of my fishing and tying strategy. You must learn *not* to judge a Klinkhåmer by looking at it in a fly box or in your hand but *only* when it floats in the surface.

The hackles on my Klinkhåmers are not tied very beautifully either because most often the fish prefer to take the most shaggy or bushy ones. The best proof of this is the fact that destroyed or heavily used flies still catch a lot of fish!

Many questions I get are about hackling and how many wraps I do for my Klinkhåmers, and I am very sorry to tell you again that it only comes with experience and lot of practice. But let me help you a little. As you wrap your hackle you should begin to learn to 'feel' what is required. There is no short cut. I actually allow the feather to 'tell me' when to stop. I learned this amazing method from Master flytyer Wayne Luallen from the United States. Not that I needed the explanation for myself, because I already discovered it through touch, but it turns out to be an unbelievable help when I try to explain to people my number of wraps. Wayne reassured me that as I wound the hackle down the wing

Nicely tied hackle and full thorax viewed from underneath

A famous shot from the 1980s on how the Klinkhåmer Special floats on its horizontal hackle

base and reached the bottom of the base, I would observe and 'feel' how the barbs suddenly start to turn upward. When this occurs, and if you then unwind the previous wrap and try to redo it, you will find this 'upward turning' will recur. This is the signal to you that the available space on the wing base has been filled with the previous windings and the fly is 'telling you' that it has had enough. I observe many people winding their hackle in the opposite direction, that is, working up the wing. But this guarantees the hackle will be less durable and eventually may come off suddenly. If you work wrapping from the top of the wing base down, it will ensure a compact well-compressed hackle and will give you the most durable construction.

Now let me explain another important technique that I use continuously when tying my parachute flies. This also will answer your question about which side of the hackle is the best to be tied up or down. It's a personal choice. I know quite a few people who tie parachute hackles the opposite way to me just because the horizontal hackle simply looks better. I don't mind less beautiful hackles if it means my catches are better. We all know that the shiny side of hackles is the one that gives birds the best protection against wet weather, and that means a *tiny* and slim parachute fly will always float better when it floats on the shiny side of the hackle. But floatant also works better on the shiny side too. So I teach my students to always tie a parachute hackle with the shiny side down in order that they become familiar with the technique in case they want to tie small or slim parachute flies later.

For the smaller ones, it is important that the fly floats on the shiny side of the horizontal hackle. There are quite a few benefits to using this technique. Tying parachute hackles with the shiny side down doesn't just give better colour, but it also curves the hackle barbs upward, because the feather has a natural curve of the barbs curling toward the bird's skin. Just try it and notice how the barbs protrude upward when the shiny side is down. I love that upward (over) curving because when flies float on the hackle, they tend to go deeper – and hanging deeper means catching more fish. I want to keep the

Ina with a nice grilse caught with a black Klinkhåmer variation in Newfoundland and Labrador, Canada

Lenok trout seduced by a large size 4 Klinkhåmer in Mongolia

body of my parachute flies as deep in the surface film as possible. Curving hackle fibres upwards also gives a better marking on mottled hackles, since the mottling does not go through to the back of the hackle very well. It also takes less hackle to do the job.

Please notice that when using my hackling technique, I FIRST tie the hackle in with the shiny side up and then bend the hackle as soon as I start doing my hackling wraps. This bending ensures that the shiny side is always downwards.

For ten years I also used a lot Coq the Leon saddle hackles for my large Klinkhåmers, Danica and Vulgata parachute flies, because they offer exactly what I wish most.

The wing and wing post

This is probably the most important part of this story. If my long-distance eyesight had been much better, I would have used Tomas Olsson's melted wing technique for most of my parachutes and never tied a Klinkhåmer. But when I started to produce parachute flies, my first priority was to be able to see and locate the fly easily in broken water. After bad experiences with calf tail as wing, I searched for a wing material that was lighter than water, not too bulky when tying it on the hook shank, not too slippery, and soft enough to allow me to pull the hackle stem deep into the material with each wrap. Finding the best material was not difficult thanks to my earlier tying experiences with the Rackelhanen, and playing so much with sedge wings made from poly yarn.

When you tie a lot of parachute flies, you quickly realise that the wing, hackle, and tying thread must cooperate in perfect harmony. If the thread or hackle slips off the wing too easily, the parachute won't last very long. The biggest mistake that people make when trying to copy the Klinkhåmer Special is not using the correct yarn for the wing. People *must not* tie Klinkhåmers using wing material that is heavier than water and they *must* use yarn that absorbs water! I showed them their mistakes when I did a yarn test in a dish of water. If you use big (heavy) hooks like me, you can't float a fly long when the wing and dubbing are heavier then water. During my workshops I advise tyers to check the yarn before they buy it. Just clip off a tiny piece of yarn, roll it into a little ball, and drop it into a dish of water.

My favourite wing length which I mostly adjust in size at the riverbank if need be

First check if it absorbs water as soon you drop it in, and how fast the absorbing process goes. The faster the better! Because of the absorbing ability of poly yarn, it will sink through the water film, but it remains floating IN the surface film forever. If you pull it down, it bobs up again because the fibres have a specific gravity of .94 and therefore should float. If it sinks to the bottom, don't buy it because my thoughts behind the fly will not work.

When you have done those two tests, cut off a new piece of yarn and make a new pellet, but this time put some floatant on your fingers and let the poly yarn absorb it which makes the yarn almost waterproof. Now drop this pellet in the water and you will see that it stays ON the surface and doesn't absorb the water at all. If you have the right floatant, the fly keeps floating for a whole day easily. You must use a floatant that is quite thin like a liquid. Some really good floatant can turn into gel when it gets colder, so I store it close to my body while fishing in cold regions. To attain the best of the best, you have to warm it up first until it liquidises and the yarn can absorb it well before the fishing. Many people are crazy about Antron yarn but I tell you it doesn't do the same as poly yarn. Antron is nylon and has a specific gravity of about 1.12 what means its heavier than water. Also it is less absorbant than poly yarn because it has a denser molecular structure.

Semperfli and Wapsi both have good quality poly yarn and are ideal for making big wings because of the strand thickness. Some of Semperfli's poly strands are a bit thinner and therefore better suited for wings on normal and smaller Klinkhåmers and parachute flies. Both are the same material. I use big bulk spools from Italy and although they are very old, they are filled with extremely good quality poly yarn. The thicker strands are superb for my large flies and I have used it since the 1980s. I prefer the thinner Semperfli yarn for my normal and smaller parachute flies. The strand thickness is extremely

Poly yarn on bulk spools which I have used since the 1980s

A beautiful Mongolian Lenok trout grabbed a huge size 4 Klinkhåmer Special

important in the tying process but I made my own rules while using it. I use two strands of Wapsi yarn on size 12 and 10 hooks. Two strands Semperfli yarn I use on size 12-14-16, and I use one strand of Semperfli on size 16-18-20.

You will have noticed that I use one and two thick strands for size 12 and it is easy to explain the reason. If I want to use the size 12 hook in more turbulent water I grab the one with the two-strand wing. To attain the same amount of wing material as Wapsi with Semperfli yarn you need to double or triple the amount – just check and practise. In this chapter I will talk about the strand thickness that is equal to Wapsi's to prevent too much confusion, even though I mainly use Semperfli's nowadays. The strand thickness is easy to adjust by yourself by taking off or adding fibres. Three strands I use for size 8-6, four strands for size 4; and six strands for size 2 hooks. That means that you will need 12 strands of Semperfli yarn for a size 2 hook to make up the same amount.

I use whatever wing colours allow me to see the fly best on the water. Usually this means a white wing, but sometimes a fluorescent or dark-coloured wing post stands out against the glare on the surface. Over the last ten years I have also successfully matched the wing, hackle, and body colours to imitate local insects better. Poly yarn is available in dozens of colours and that's probably why there are so many Klinkhåmers with so many coloured wing posts.

When I started fishing with the Klinkhåmer Special, the large visible wing was invaluable for catching grayling that suck the flies from the surface without leaving a trace. The same happened with Atlantic salmon, and some other species. As soon as the big white wing disappears, I trust my floatant and set the hook knowing a fish must have taken the fly. I land countless fish that way.

I also make a series of Klinkhåmers to use in very turbulent waters, for which you must find the perfect balance within your fly. To test the balance, I refer you again to your dish with water because that will be your best friend to find the magic touch. Wild water Klinkhåmers as I call them need much more wing material than the normal ones. Always keep in mind: with double or even triple the strands of poly yarn, the hackle is not really so important anymore for a proper floatation. The poly yarn becomes the only element that keeps the fly floating.

So to find the correct balance in your flies, just add floatant to your wing and test your completed fly in a dish of water. In the beginning I even created turbulent water myself to try to get the fly to sink. Many did and so I kept adding strands of poly yarn until the fly floated. That's why I often have much thicker wings than everybody else on my Klinkhåmer Specials for turbulent waters. I use the same trick to work out the amount of poly yarn to use on really big hook sizes. I achieve success by constantly using my dish of water. When I do quality tests for commercially-tied Klinkhåmers, a dish of water is also indispensable. For instance, I was testing a few dozen Klinkhåmers on their quality and all sank even after I had put floatant on the wing. The maker had used superglue to secure the hackle better but the superglue hardens the yarn and prevents it from absorbing the floatant. So they didn't work. This is why I use a small eyebrow comb, as soon the fly is completed, just to be sure all fibres in the wing are separated, and that trick lets the yarn absorb the floatant even better.

I have tried using other materials such as CDC or calf tail but both get waterlogged after playing fish for a while. I used foam with some success, but I always go back to yarn. Foam will help the fly float even better, but I find the yarn more natural and easier to cast. If you use poor quality foam, fish teeth can easily damage the wing as well. Quite a few people use foam for parachutes because if you cut the foam directly above the hackle, the foam expands and pushes the hackle fibres downward, which often gives

Giant unsinkable Klinkhåmers with enormous wings from poly yarn developed in Malaysia several years ago

a really beautiful appearance to the hackle too. Foam helps the fly to look and float better but I never caught more fish with foam posts than with poly yarn wings; always the opposite.

One final comment about the length of the wing: I see many Klinkhåmers with a poly yarn wing post which has been cut much too short. The wing should be at least the length of the hook shank or longer. A wing that is long can easily be shortened later, but it can never be lengthened. Long wings also give larger silhouettes and in some weather conditions I have caught more fish using a Klinkhåmer Special tied with what many people consider to be an extravagant wing.

In the past, I stored all my Klinkhåmers up to size 8 in Wheatley 6-inch extra deep fly boxes. The storage of parachute flies and flies that have long wing posts is crucial for me. I don't like it when wings on my flies get reshaped or damaged by storage so I use the extra deep boxes to give them space for wing and hackle. Nowadays, I only use the big waterproof boxes from C&F with the exchangeable foam patches. These boxes are very deep and perfect for my big long-winged Klinkhåmers. For sizes 6-4-2 and 1, I just use an extra deep big plastic compartment box.

Note: Doug Swisher and Carl Richards made the first mention of poly fibre in their book *Selective Trout* (1971). They felt its specific gravity was extremely important for dry flies. It is a consideration that has been debated many times, and may or may not be true. The book is worth a read.

A fantastic wing update

When I finally retired in 2012 and had the unique opportunity to study and photograph the *Ephemera danica* in Denmark for five weeks in a row, I entered a new world. As a member of the Danish Sport Fishing Organisation I was allowed to fish several rivers with great hatches of mayflies and countless caddis species. For the first time in my life I was able to follow the hatches upstream and could take the most

My own mixture made from five different colours of poly yarn. Just pick a few strands of each colour and roll them by hand on a hard surface. Klinks made with these wings, I see as version 3.0

My special wing colour for a perfect match the hatch; my variation I call Klinkhåmer 3.0

The Klinks 3.0 that I use during mayfly (Danica) hatches with great success

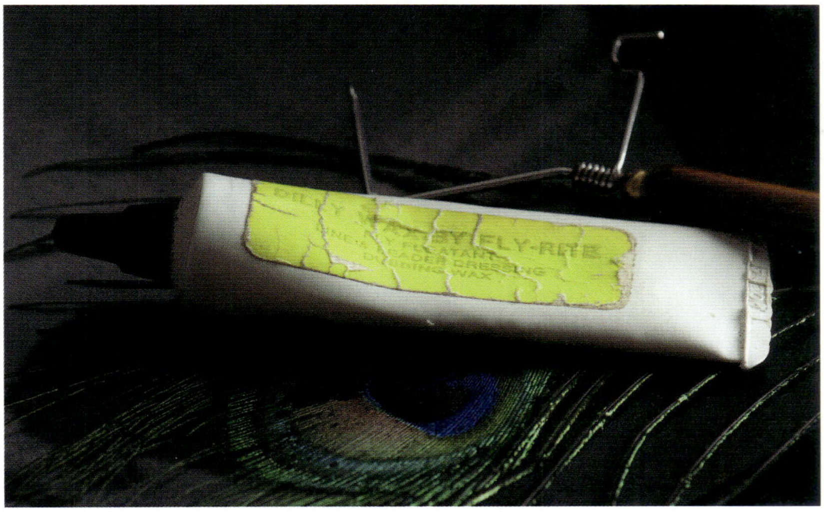

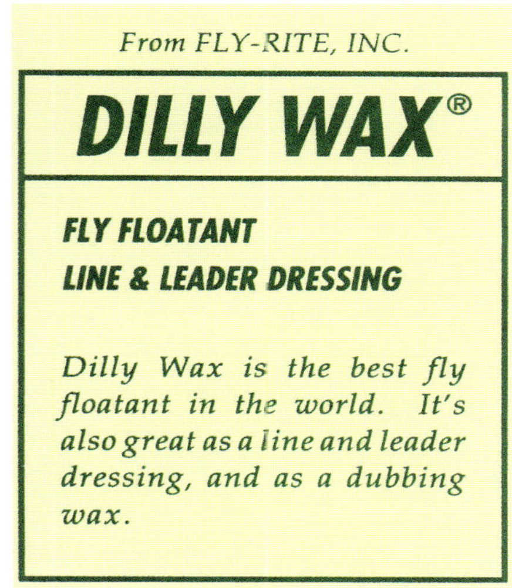

beautiful Danica and sedge pictures imaginable. I was also able to roam with my camera, tripod and macro lenses for more than 30 days. I had never seen so many Danicas in my entire life and took over a thousand photographs, pictures that were desperately missing in my nature and wildlife database.

I learned a lot and studied more than I fished. When I had opened my fly boxes on the riverbank and the Danicas landed in the boxes, my flytying heart was beating like crazy as I could see the real insects among my artificial Danicas. Many ideas awoke but mainly for improving the colour of the wings for my flies. I had tied the tails, wing and bodies perfectly and even in the correct size but the wing colours needed a huge and final update. During the winter, I worked for weeks to try to find a good solution. I finally succeeded by mixing it myself.

To imitate the Danica and Vulgata wings I created my own wing colour of poly yarn by mixing brown, black, cream, grey and transparent poly yarn. The final outcome was overwhelming, resulting in what I believe is a great wing colour for many mayfly species. I proved that by catching far more fish than before. It is a lot of work to make your own wing colour but worth every minute! This is why I also make so many of my own dubbing materials too. It's all part of flytying for me.

Dilly wax

Some people have problems keeping their Klinkhåmers floating. The main reason for this is the use of poly yarn for the wing, and I have explained most solutions in the previous paragraphs. Poly yarn is indeed much lighter than water but after it has absorbed water the weight of the hook will quickly pull the fly down. The use of a good floatant is essential to prevent this problem. Therefore the wing should be treated with a very good floatant *before* you use your flies. Also apply floatant to the hackles of your slim and normal-sized Klinkhåmers and put it on the tail of different parachute flies. I highly recommend using Dilly Wax from Fly-Rite which is specially designed to make polypropylene yarn and poly dubbing water-resistant. Dilly wax makes a well-tied Klinkhåmer almost unsinkable: just put a drop on your finger or wing and rub it in well. Properly done this keeps the fly keeps floating for a very long time.

One word of caution: Do not use floatant on the bodies (especially on small sizes), because it can prevent them from breaking through the surface of the water. This is extremely important

My two favourite body materials for Klinkhåmer Specials

to the success of this unique pattern. Dilly wax also can be used as line dressing for tippet, leader and fly line, and because it is not toxic you even can use it as lip-balm. Be careful when using it in warm countries or hot weather because Dilly wax will turn from a gel to a liquid when the temperature goes above 25 degrees.

Body materials and ribbing

I have seen many Klinkhåmers with enough dubbing for four or five flies. Believe me, slimmer bodies produce more fish, so a sparse use of dubbing material is essential. I mainly use four materials as body for most of my Klinkhåmers. For normal and large flies, I have found nothing better than a body made from Fly-Rite poly dubbing. The solid and perfectly blended colours (46 in total) will help you to imitate any insect body colour anywhere on the globe. The original colour for the Klinkhåmer Special is light tan (number 19) but I also frequently used dark tan (number 20) because I found both colours most useful for the Scandinavian countries and subarctic regions all over the planet. I use the Borger Color System booklet (*see later*) to colour-match my Fly-Rite poly dubbing perfectly. With the BCS booklet it is very easy to pick the right shade of green from its over 40 colour swatches of different shades of green. By using the same system, everyone talks the same language. I love the BCS, especially when I do my insect research and tying afterwards. It's a pity not more people use it because it is very handy. Let me give you a simple example: somewhere in the 1980s we discovered an insect with a body colour that perfectly matched the BCS colour code 101. After some research and tests, we were able to make an exact body colour by mixing Fly-Rite dubbings no. 42, 43 and 32. The catch results were amazing. Here is another great insect colour I found in BCS booklet, page 8 under colour code 29: I mixed Fly-Rite dubbing no. 41, 42 and 13.

When I first started to travel to places with totally different insect life, I changed the body colours more frequently. You can blend your own colours very easily with Fly-Rite dubbing which makes 'match the hatch' tying and fishing a lot simpler.

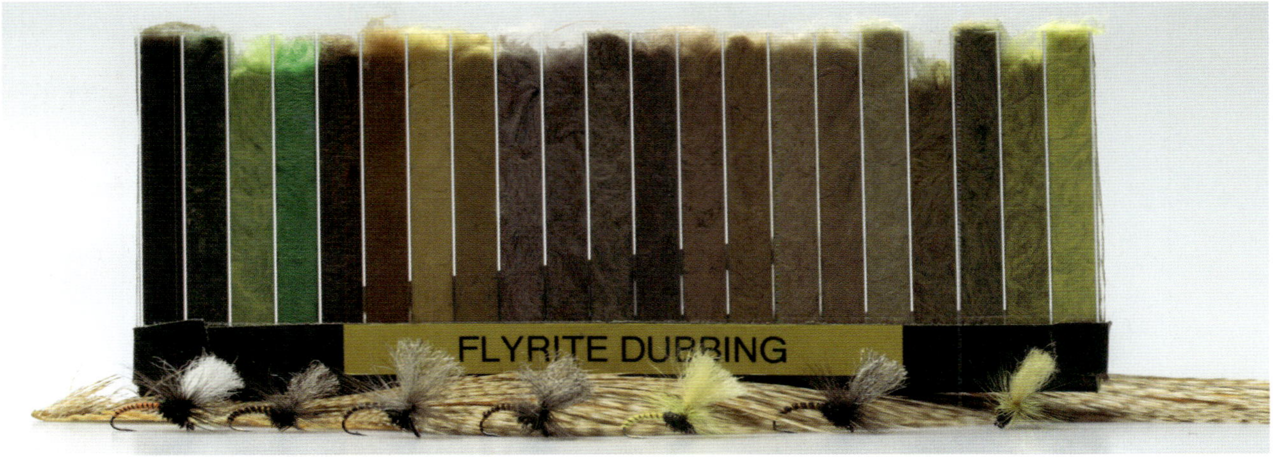

A small selection of Fly-Rite dubbing in my own storage dispenser that I usually take with me

Semperfli's pure silk thread that I use for ribbing in all my quill body flies these days

For slim and smaller flies, I often use peacock quills or a few fibres that I carefully cut from quills from large natural wing or tail feathers. When using quills as body material I much prefer them in natural black and when you rib it with yellow or classic orange (6A) pure silk, it becomes a deadly combination.

My third body material are strong hairs to create small but durable bodies on my Klinkhåmer flies, for example a peccary hair or similar-looking fibre.

The fourth and last material that I use intensively for my Klinkhåmer bodies is Kapok dubbing. For more than 15 years it was my secret dubbing. I got it directly from Indonesia but I could never share it with the world because my supply was very limited. What I received was just enough to make around 700-800 flies on a yearly basis. I only had it in four colours and every time my annual package arrived, the colours were always different.

When you talk or write about colours there is nothing worse than an unpredictable colour range. I only saw Kapok dubbing as experimental body material but with amazing and incredible floating abilities. I really loved it, so you can't believe how happy I was when Semperfli was finally able to supply Kapok in huge amounts and in 28 different shades of very reliable and awesome colours.

The Kapok tree (*Ceiba pentandra*) has been available for many years growing throughout the tropics where it is cultivated for its fibres. Kapok is a moisture-resistant, quick-drying, resilient, and buoyant fibre which weighs only one-eighth of cotton. Kapok has been used in life jackets and life boats, supporting as much as 30 times its own weight in water. Kapok buoyancy is lost very slowly.

Historically Kapok is very difficult to dye, but the team at Semperfli have been working with computer-based dying machinery which has enabled it to bulk dye Kapok with replicated colours every time. Together with Semperfli's Pro Team they defined a good range of very natural dubbing colours to produce some of the best dry fly dubbing available in the market today. You can buy it in individual colours but also in a Natural Dispenser in 12 colours, a Mayfly Dispenser in 12 colours and a full Kapok Range Dispenser in 22 colours.

A closer look at Gary Borger's Color System

To select a correct body colour easily, work a lot with Gary Borger's Color System. You have to appreciate that it is no use picking and selecting dry dubbing materials: that is a very common mistake. When I use the BCS system I only use it to analyse *wet* colours when it comes to body colours.

Fly-Rite dubbing no 19, 20 and 21 were my most used colours in the 1980s and 1990s

A precise colour description by phone is probably one of the most difficult things for a flyfisher to explain. In my early days, I mostly used the dubbing numbers from Fly-Rite to match the colour as closely as possible, and only selected it wet first. Since Gary Borger introduced his BCS, the communication of colours became a simple, accurate procedure. The colour swatches inside the booklet were selected on the basis of twenty years of research and in my circle of friends, we use the BCS to prevent colour misunderstandings. I don't know why so few people in Europe use this excellent colour system. Why not use a system that has proved its power over so many years? I have used it exclusively since 1986.

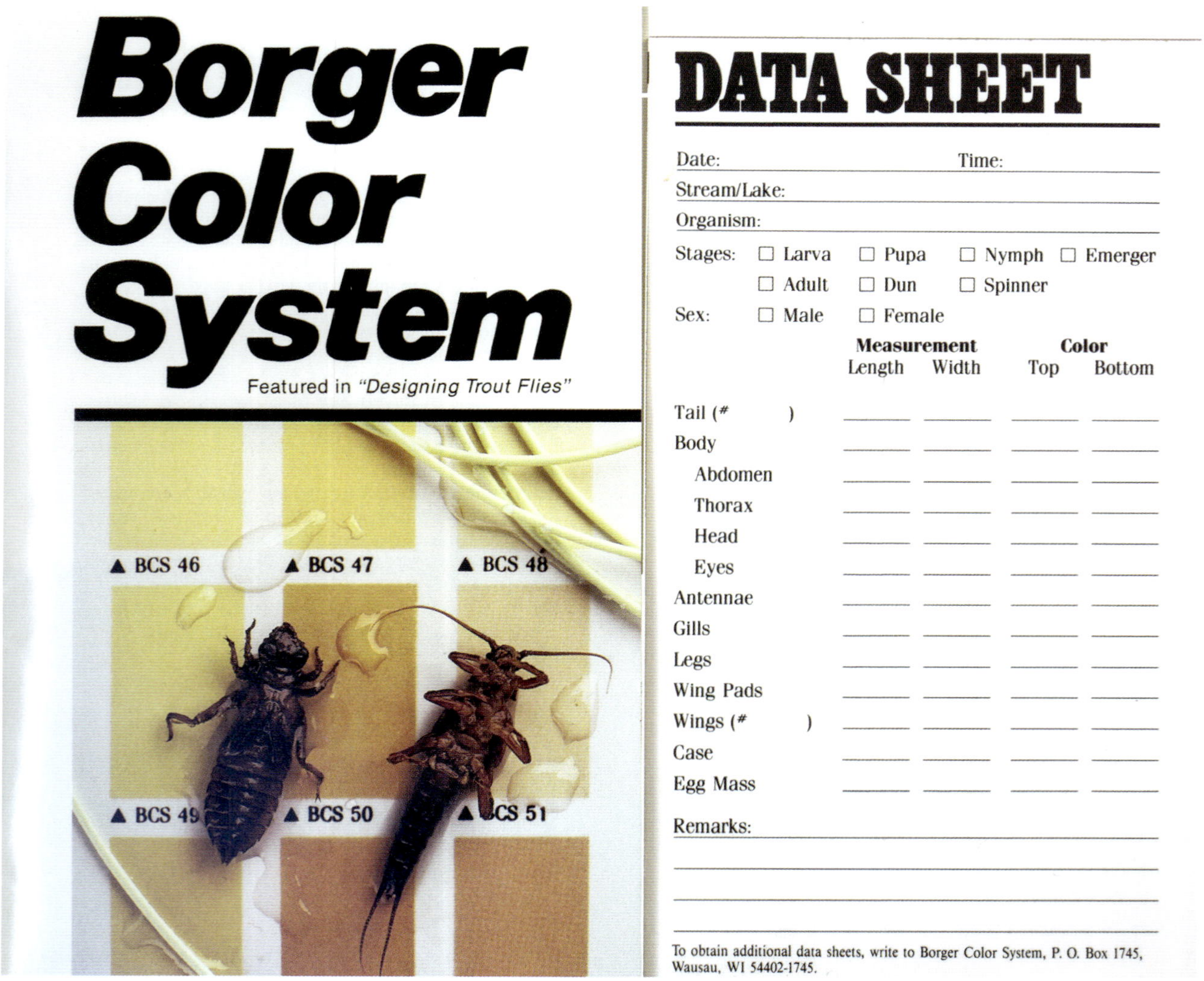

Garry Borger Color System which I still use today with great pleasure

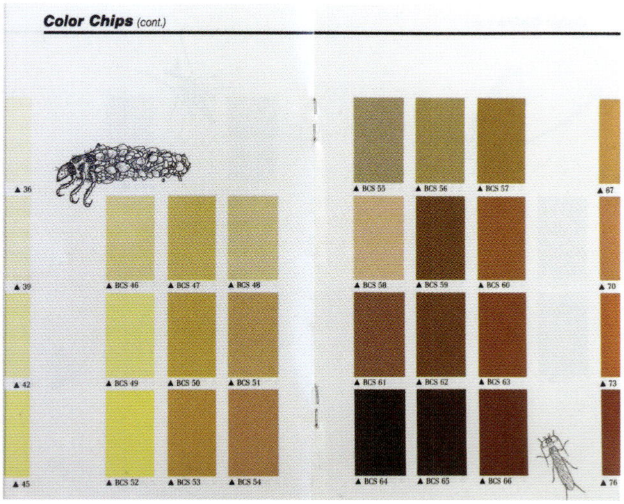

Set of my favourite kapok colours for Denmark.
Colours top row: Danica, Tan, Adams
Colours bottom row: Brown and BWO

Kapok dispensers filled with great natural colours of the real insects

Left: The colour swatches inside the Borger Color Sytem booklet

The BCS is a small booklet that fits in your fishing vest. The colour panels of the highest quality of waterproof inks are printed on durable, water-resistant paper. An additional data notebook helps you to make a useful observation list from the insect species you collect from your favourite waters.

To work with the booklet is very simple. Inside is a well-described user's manual. It becomes a matter of collecting your specimen and putting it beside the colour panel that corresponds most accurately with the colour of the insect. Fill in your data sheet, and you can always make a perfect copy later.

The thorax

One of the most important elements of the Klinkhåmer Special is the noticeably enlarged thorax made from peacock herl. I love a big thorax on my Klinkhåmers. Peacock herl is notoriously fragile but you can twist it around your tying thread to improve its durability. Oliver Edwards explained in his workshops how he ties in the peacock herl at the tips to make the tying easier. His method produces a nicer looking thorax. I personally strengthen the thorax by pressing a drop of special varnish into the base of the wing with a varnish applicator, in order to secure the wing, thread, and thorax after I have completed the fly. Depending on the hook size I take 1-3 herl of peacock to form the thorax.

Over the last 15 years, synthetic material has become more and more popular for creating powerful thoraxes. Semperfli's Straggle String is a unique and specially formulated micro chenille that is ideal for Klinkhåmer thoraxes but you also can use it for bodies and to create extra legs on your nymphs as well. I use the fluorescent colours exclusively for creating the famous IJssel

Left: Quills and fibres which I also use regularly for my Klinkhåmer bodies

The Klinkhåmer registration system

All the Klinkhåmers that I made between 1984 and 1991 are stored in this particular box

The contents of this box are flies made only with Fly-Rite bodies

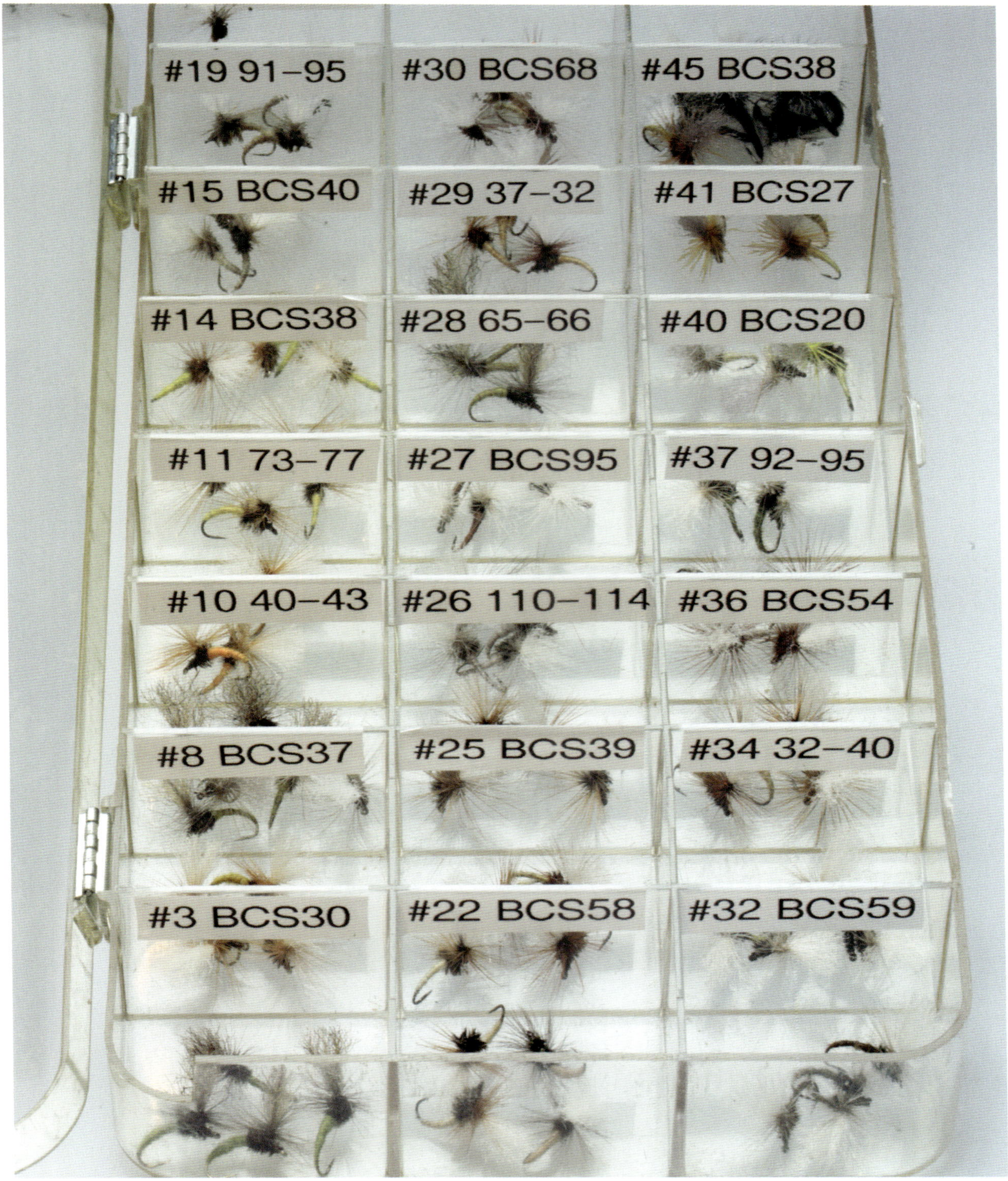

*Box one from my private collection of my Klinkhåmer flies. This is just one of the five boxes that I have today. In this box are all my flies that I have registered since 1984, made with bodies in the Fly-Rite colours. I made four of each fly, just in case I lost one or two. Every fly is registered with the body colour of Fly-Rite that I use, which matches almost 100% to the BCS colours I determine from the insects that I found in the water or riverbanks between 1984 and 2024. From each fly registration, I also have a list of the rivers and lakes where these patterns were most successful. For example, the fly in box **#8 BCS 37** is made from Fly-Rite colour no. 8 and it imitates an insect with body colour BCS 37 on Gary Borger's colour sheet. The fly was most successful in the River Nora in eastern Norway between 1984-1988. Thanks to this system I was able to give really good advice to my friends and club members of the Grayling Society Holland who wish to fish in Scandinavia*

and Ide nymphs for Holland. Straggle legs, also micro chenille, is another perfect synthetic product from Semperfli that I use a lot for my thoraxes. Because the micro legs are a bit longer than Straggle string, I use the Straggle legs for bigger Klinkhåmers.

A brand-new product from Semperfli that first came on the market in December 2023 is the black, green and blue Synthetic Peacock herl. I was in the lucky position to test the 2mm strands in Finland and Denmark with unbelievable success. In many situations, it even was better than original peacock herls. The shiny colour of the strands is amazing and it will now be the best replacement for real peacock in most dressings. After the tying steps of my original fly, I will show you this substitute for peacock herl in my updated version of my Klinkhåmer Special (version 2.0) and will explain why I use it and how I tie it.

The truth about hooks

I know everybody has their own preference for hooks; my favourite trout and grayling hook is one with a nice wide gape, long curved shank, straight eye, mini barb and fine wire. Yes, you read it correctly: I still want a barb, despite the fact that I fish barbless all the time. I prefer a mini barb for two reasons: firstly, because I always lose flies in my fly boxes when I have pinched off the barbs before the fishing, and secondly, when I want a fresh fly for a huge fish that I want to catch, I know that particular fly wasn't fished before because the barb is still on. My first Klinkhåmer Special was tied on the Partridge K2B, better known as the 'Yorkshire Sedge hook', and at that time it was available in very fine wire. I didn't like the up-eye, but I chose that particular hook because it had the exact size and curve as the caddis larvae I had found in the stomach of that famous grayling.

My knowledge of entomology wasn't spectacular in those days. One of my earliest strategies was to measure real insects on hooks to find the best hook size to 'match the hatch'. However, when Partridge changed the K2B hook – and started using heavier wire – I searched for something else. For a while I used the Partridge grub and shrimp hook known as the K4A because of its down-turned neat ball eye. It was a little easier with the parachute, but I was not fully satisfied with it. I still wanted that hook with a longer shank, a very fine wire and straight-eye. A straight-eye was very important to me because it was not only easier for tying parachute flies, but I could also use the same hook for other emerger patterns and even nymphs. I loved hooks with multiple purposes in the very early stage of my flytying development; maybe that's why I have always tied so many different patterns on the same types of hooks.

However, when I wasn't able to find the hook that I needed so desperately, I started to reshape the Partridge K12ST and later the GRS12ST

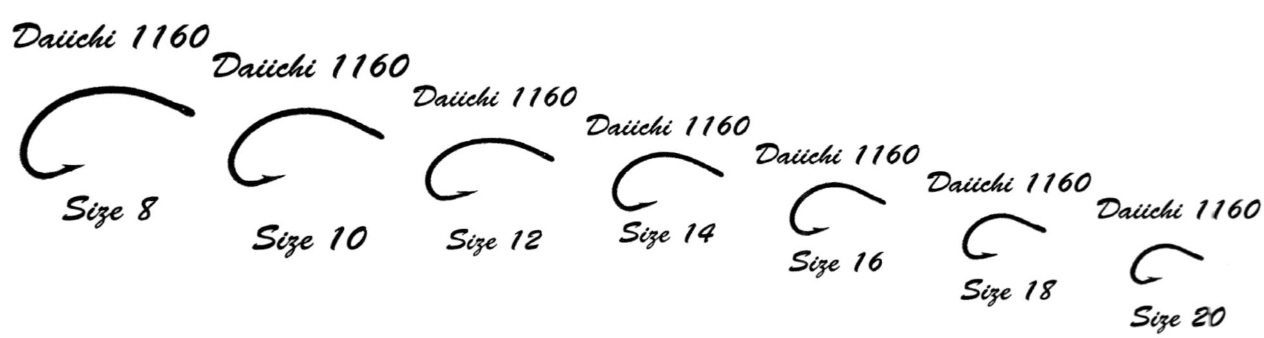

The range of Klinkhamer hooks size 8 down to 20

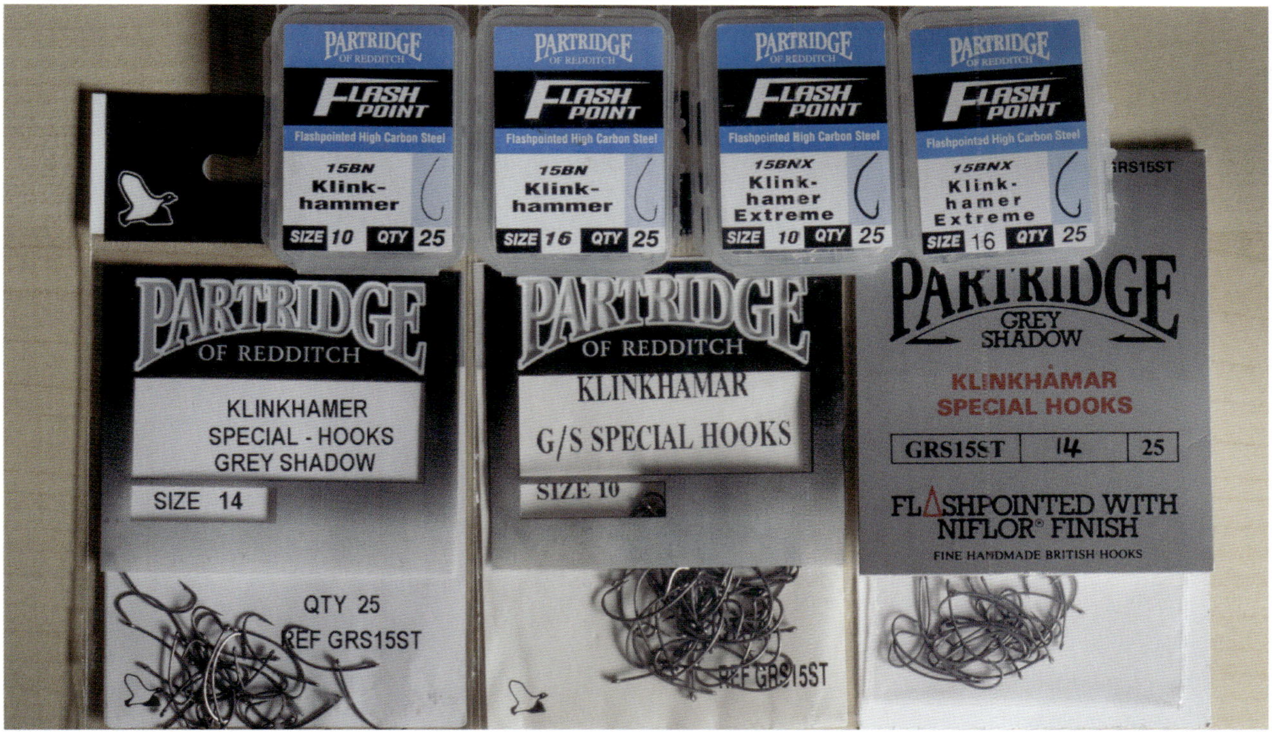

Just a sample of all the different names for my hook: no wonder there was always so much confusion

hooks to create a shape that I wanted for tying my Klinkhåmers. It was not good publicity for the Partridge Hook Company because I broke so many during the reshaping process. I finally designed my own hook for Alan Bramley, who had owned the Partridge Hook Company in Redditch since the 1970s. We'd been friends since the early 1980s, but we became very close after I had developed my Klinkhåmer.

Because of the reshaping problem, Alan had given me some soft hook wire so that I could show him this special hook shape I was talking about all the time. With this special wire it was very easy for me to bend and shape a perfect example of the hook that I wanted so eagerly. At first Alan was a little sceptical because of the large size, but his adviser Terry Ruane really liked my idea and so they started making some prototypes. I think it was the fourth prototype that I was really crazy about. It was a black version and for quite a while it was for me the most beautiful grayling hook ever designed. Sadly it was a hook to catch grayling only, and I have to be honest because the hook was so light and thin that it often broke when you hooked a large trout, so it was actually never produced in large quantities. There have been a few other small productions of varied prototypes since the late 1980s, but none of them really made it commercially despite the fact I used some of them with great success. There was a K15ST Klinkhåmer hook for a short while, but the Grey Shadow hooks with points that were chemically sharpened with a new method called 'flash pointing' became really popular in the early 1990s. The new GRS hooks were extremely strong and sharp, but because they had a nilflor finish, the surface was very slippery too. A lot of flytyers (including me) had to get used to that at first. My hook finally found its way as the Partridge Klinkhåmer hook GRS15ST, and it became one of the bestselling hooks in Partridge's range for quite a while. The hook was renamed and redesigned as the 15BN and 15BNX when Mustad took over Partridge of Redditch in 1996. Sadly, the 15BN and 15BNX have changed from the original shape and wire over the years.

I was very loyal to Partridge for 33 years until Fishing Matters Ltd bought Partridge in 2009, and

The new Daiichi hooks, perfectly made following the original blueprint from the mid 1980s

the Klinkhåmer hooks started to become in short supply. There was no stock for quite a while, and when I had no hooks left to continue my workshops and classes, I knew I had to do something. I started to collaborate with Anglers Sport Group in the USA. ASG distributes Daiichi hooks and Daiichi was already famous for manufacturing some of the best hooks in the world.

The Daiichi owners really liked my early hook design and the flytying projects I was involved with. After some serious talks and when they were willing to sponsor my flytying programmes for kids all over the world, we quickly came to an agreement. A good hook sponsor was very important for me, so I could continue my kid classes and workshops free of charge. When we set up our new partnership, we agreed that the Daiichi Klinkhåmer hooks would be made 100% to my exact design from the 1980s. With a new start I also was able to put the record straight and name the hook properly. This was very important for me after Partridge and Mustad had used six different names for the hook.

In amazingly short order, the first prototypes of the new Klinkhåmer hook passed all my field tests and were an exact copy of my original design. They are on the market in sizes 20 to 8 in two versions: bronze (Daiichi 1160) and black nickel (Daiichi 1167).

The new hook is most suitable for tying Klinkhåmers, emergers, and many special nymph patterns. Daiichi originally planned to make them in smaller sizes but I personally didn't need or want them smaller than size 20 so they dropped the idea. Instead I would prefer them also in sizes 6, 4 and 2 but the market was not ready for such a big size. When fishing in Arctic regions in June and early July I don't use sizes smaller than 14. In Lapland and countries like Alaska, northern Canada, Russia, Mongolia, and for Atlantic salmon fishing, I mainly use private Klinkhåmer hooks in sizes 2-4-6 as leftovers from earlier prototypes.

People who want to fish with big dry flies for really big fish like pacu, catfish, jungle perch and Sultan fish, I usually advise to use the Daiichi 3847 circle hook in sizes 1 and 1/0 or use the beautiful curved saltwater hook in size 2 from Grip (44378TEF) but then take 5-8 strands of poly yarn for the wing.

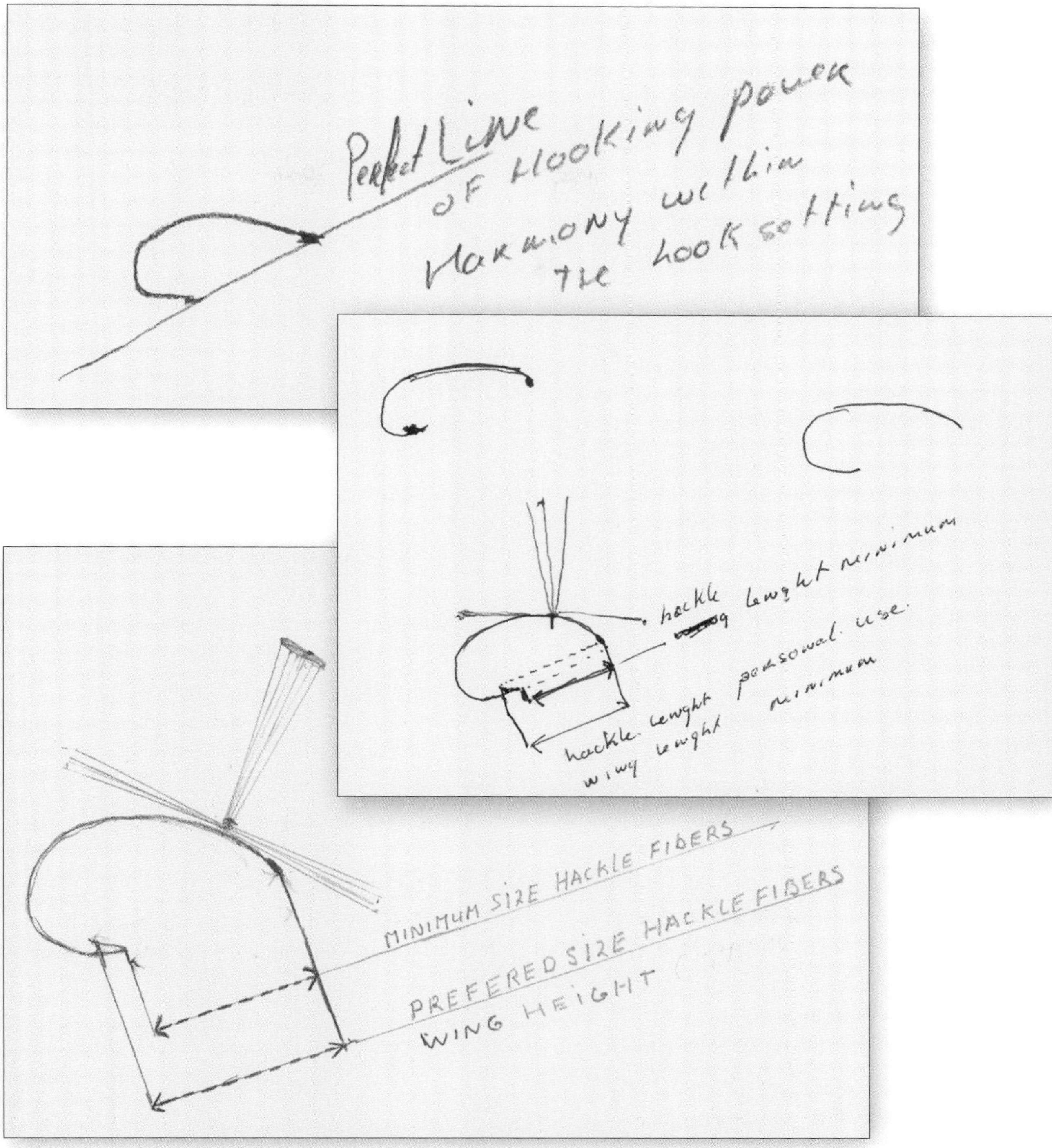

A few of my original Klinkhåmer hook sketches from the 1980s which I drew sitting at the kitchen table with the late Peter Rosmalen

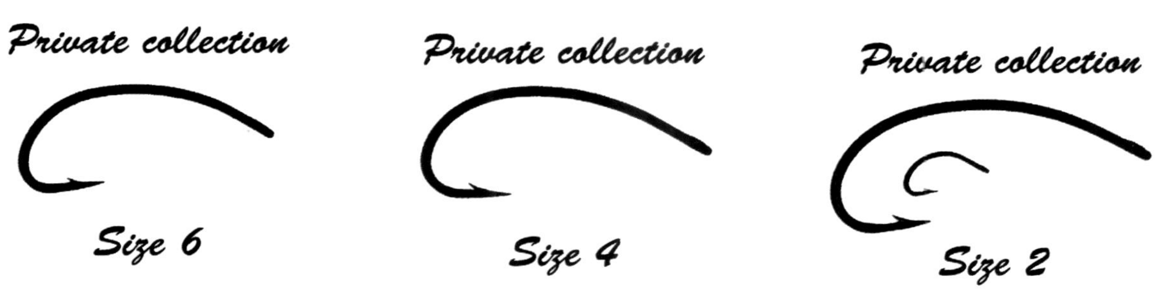

Above: my private selection of large Klinkhåmer hooks from the early 1990s. The little hook is size 20

Size matters

Over the years of using the fly, my Klinkhåmer hook sizes have grown enormously and the biggest size I tie today is size 1. The problem, however, is that not many hooks have larger long shank curved hooks than a size 8. I am very fortunate that Alan Bramley made me 1000 Klinkhåmer hooks each in sizes 6-4-2 and 500 in size 1, and gave them to me as a very special birthday present in the early 1990s. With those big hooks suddenly available, I was able to experiment at once. I made some steelhead Klinkhåmers for friends who often fished British Columbia and Alaska and designed my own large variations to use for Atlantic salmon. I always had a few big sizes in a separate fly box and it wasn't long before I tried a really big size for grayling and trout in Norway, Sweden and Finland too.

It worked amazingly well and very soon a complete new fishing strategy was born. **Today I always start with my biggest Klinkhåmer, wherever I fish** then I go down in size until I start to catch some fish. Often big fish have already grabbed one of my big sizes before I was able to go smaller. In areas with short summers such as northern Scandinavia, Alaska, Russia, northern Canada, Mongolia, Chile and Argentina, this strategy works awesomely. If you do it the opposite way, this trick doesn't work because I already told you that small Klinkhåmers catch a lot of smaller fish. Once I started to fish frequently in countries like Malaysia or was able to experiment in rivers in South Africa, I discovered that many other species seem to like giant Klinkhåmers as well. I also learned that it is possible to hook large fish like catfish, pacu, carp, jungle perch, taimen and even sultan fish with enormous Klinkhåmer flies.

Hooking problems

The most common mistake that people make is they don't use the right hooks for their Klinkhåmers. Many people write to tell me they have problems hooking fish with a Klinkhåmer Special. They get many strikes but still miss a lot of fish. When I ask what kind of hook they use, it usually turns out they are using smaller nymph, scud or caddis hooks, curved but with a short shank and much smaller gape compared to the special hook shape that I so highly recommend. When you use a small hook, or a hook with a small gape, you dramatically reduce your hooking potential.

Spiderweb* and Nano silk

(SPIDERWEB see notes page 153*)

The greatest problem with my first parachute flies was that I was not satisfied with the durability of the parachute hackles. Although I fished intensively for grayling, I occasionally hooked a trout. Those trout often destroyed my parachutes, and from

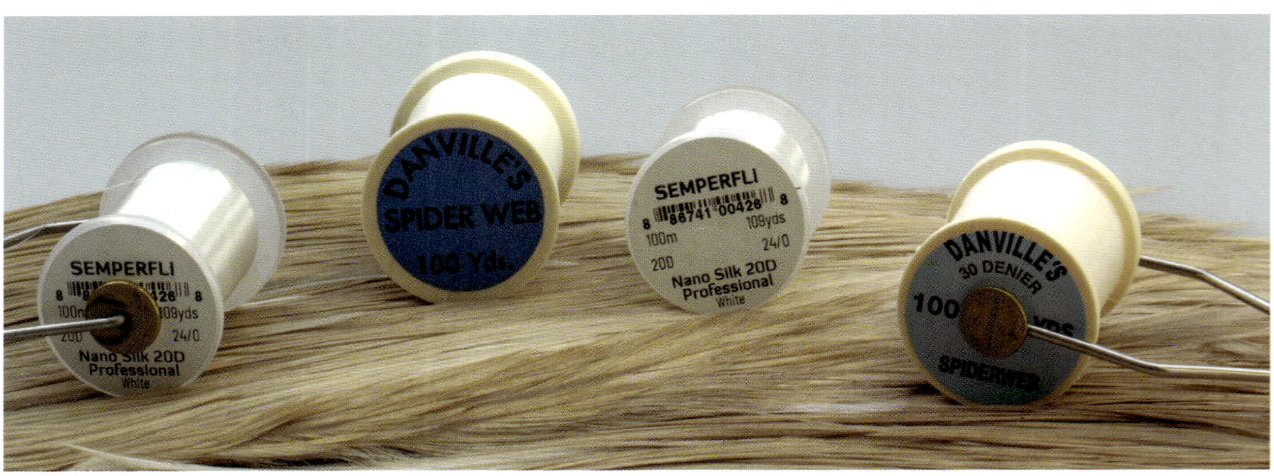

Spiderweb and Nano Silk 24/0 are the only two threads I use for my special technique to secure the hackle

Amazing that this small grayling took a size 4 Klinkhåmer in Mongolia

that moment, making more durable parachute hackles became my highest priority. I have no idea why so many people still tie off a parachute hackle at the eye of the hook. If you do that, then tie off with a whip finisher, you first have to pull away the hackle fibres before you can secure the hackle, which makes tying off the feather more difficult. Maybe that's why so many people don't like to tie parachute flies. It is a complicated technique and it also leads to fragile parachute hackles because the hackle goes directly from the wing post to the hook eye which leaves the quill exposed to the sharp teeth of a trout or other species. When the quill breaks, the parachute hackle comes off. It took me an entire winter to find a technique that protects the hackle against sharp teeth.

I tried to do some extra wraps through the thorax before securing the hackle at the hook eye; this indeed made the hackle more durable, but it also makes the tying process even more complicated. After tying thousands of parachute flies in my life, I can say with confidence that it is much simpler to wrap a hackle around the base of the wing post, and secure the hackle tip at the base of the wing post. When the hackle is pulled deep into the wing post, and the hackle tip is tied off with several layers of Spiderweb thread, the quill becomes well protected in the bulk of the wing.

To perform this different technique of mine, I not only needed the thinnest thread available on the market, but I also needed a thread that was a bit slippery and was made from one single strand and extremely thin. Then it would go smoothly between the hackle fibres instead of grabbing them and damaging the fibres, as my ordinary thread mostly did. This finally led me to Spiderweb thread from Danville, which is actually pure nylon, but it satisfied me in all ways. It is true that it is a little fragile and you have to get used to it, but once you know the trick, it is a great pleasure to work with. The spools of Spiderweb worked best in my cheapest bobbin holder. Adjusting the bobbin is the key to success. It took a long time before other thinner threads were developed so I exclusively used Danville Spiderweb for my parachute technique and smallest dry flies for a very long time and still do today. When I introduced the thread to Oliver Edwards he loved it right away and told me it was the thread he had been looking for his entire life, and not that long after I had given him a few spools, his first tiny realistic masterpieces left his vice.

Although several of my close friends use Spiderweb to make the entire fly, I only use it for making the wing base and to tie-off my parachute hackles around wing posts. I also use it for repairing parachute flies, which means I never have to throw away damaged parachute flies if the parachutes have come off. I love my technique and it makes a parachute hackle much more durable indeed, but it doesn't mean that my parachutes are perfect and never, ever come off! I fish my flies for a long time so occasionally I still lose a hackle and an easy way to repair them is a nice bonus. I will show later in this chapter how I repair my parachute when it has come off.

Danville's Spiderweb is available in 100-yard spools and in white colour only, and for me still feels the thinnest thread available on the market today. My idea for using Danville's Spiderweb to secure the parachute hackles was probably new in the mid 1980s: certainly I had never seen it before. This method makes winding a parachute a simple operation while forming an effective, durable parachute wing without damaging any of the hackle fibres, and you will never get the same results using regular threads. Kids in my flytying classes were all able to produce their first parachute fly within an hour using this technique.

White Nano silk #24/0 from Semperfli is by far the best substitute and is definitely my second choice after Spiderweb. It's very slippery too and compared to Spiderweb it is extremely strong and therefore a perfect choice for tyers who don't like Spiderweb or can't get used to it. I always use Nano silk during my flytying classes with kids because they cannot break it. After 40 years, I still tie the parachutes on my Klinkhåmer Specials in exactly the same way.

* **Spiderweb** Danville's Spiderweb is so fine and fragile that many people refuse to use it. In reality, it's mostly a matter of using a perfectly adjusted bobbin holder for your spiderweb spools. When you use Spiderweb, you always need a very loose bobbin holder or the thread will break too easily. It is the finest thread available on the market. Spiderweb is a favourite tying thread for top tyers to create midges and other very small flies. The first spool I ever found was in a little tackle shop in Sweden in the 1980s, and the owner gave me two dozen spools free of charge! He had ordered it by mistake and didn't know what to do with it. He had probably never sold any because it was so fragile. Because of its small diameter, it gets very little build up, and it disappears into your dubbing materials. It's an ideal thread to disappear between hackles without damaging them, and it will also get really deep into a wing post of poly yarn. I discovered that a combination of poly yarn and Spiderweb seems to work in perfect harmony. Spiderweb gets its strength by the number of windings that you make. It became enormously popular in Europe after my article about the Klinkhåmer Special in 1986 and the favourable words from Oliver Edwards.

Ina's largest salmon, caught with a size 8 Klinkhåmer Special in Newfoundland 2015

Important points before you start tying

• HOOKING POWER

Use long shank, wide gape hooks for more solid hook-ups. If you can't find the recommended hooks, search for hooks with the widest gape and longest shank you can find. When you can draw a straight line from the point of the hook towards the hook eye, you have the best hooking power.

• VISIBILITY

Your ability to see this pattern on the water is vital. Tie the wing long so that it can easily be spotted in fast, turbulent waters. You can always cut the wing at the riverbank. Pick the wing colour you feel most happy with.

• A STRONG, SOLID WING BASE

A strong thread base at the bottom of the wing is essential for the fly's durability, and it makes the wrapping of your parachute hackle a lot easier.

• EXTRA STRENGTH OF THE HACKLE

Secure the parachute hackle by using Spiderweb thread or Semperfli's 24/0 Nano Silk. When you tie off the hackle, use a whip finisher to make a strong and durable construction, which also improves the appearance of the finished parachute hackle.

• TURBULENT WATER

For wild rivers, increase the amount of poly yarn so the yarn will keep the fly floating when prepared well with floatant.

• DARKER THORAX

Fishing experience has taught me that a fly tied with a darker thorax catches more fish than a similar fly made with a lighter-coloured thorax.

• THIN IS IN

Tie a slim, tapered abdomen. Experiment to test if that works better for you. The most successful patterns in my fly boxes all have slender bodies.

• LOCATION OF THE WING

A lot of people have serious problems deciding where to put the wing. That's why the 15BNX was developed many years ago. What hardly anybody knew was that the extra bend in front of that *particular* hook was actually designed to ensure that the wing was perfectly positioned and proportioned. The wing should be tied up in the bend. Personally, I am very flexible about the wing location and I mostly place the wing so that it pops out in the middle of the thorax when the fly is completed. As long as the wing is not too close to the hook eye, I am happy with it.

• USE FLOATANT WISELY

Do not apply any floatant to the body but only to the wing and hackle.

• VARNISH TO SEAL THE HACKLE, SPIDERWEB AND THORAX

Extremely important to increase the durability of the ready-made parachute hackle is the use of a good quality varnish applicator. It must have a fine needle to spread the varnish on the whip-finished wraps. The needle and absorbing property of poly yarn will help you to get the lacquer deep into the exact places you want it. All you need to do is dilute the varnish to make this technique work in the most powerful way. A mixture of cellulose lacquer diluted with acetone in a 50/50 mixture is perfect for this trick, creating the most durable hackle. It also ensures the varnish will work well with most applicators and spread easily into the wing, thorax and thread wraps.

• QUILL BODIES

When I make very slim Klinkhåmer bodies from quills, fibres or hairs, I don't make an underbody. In that situation, I just split the wing and put the centre of the poly yarn under the hook and fold it upwards and secure it well on the hook shank.

• BE CREATIVE

In my dressing, you will find only the recipes for a standard and original Klinkhåmer Special. Be creative and inventive by selecting your own choice of colours to match the hatch of the insects in your favourite waters or even design your own fancies. Over the years I have tied 121 different Klinkhåmer variations since 1984.

A few variations of my Klinkhåmer Specials

Beautiful grayling, over 50cm, caught in a lovely stream in southern Finnish Lapland

One of my Danica Klinks I had given away to a fellow angler, and later found hanging in a tree

Tying the original Klinkhåmer Special

Dressing

Hook: Daiichi 1160 or 1167 Klinkhåmer hook size 8-20

Thread: Semperfli 8/0 classic waxed thread in tan colour and Spiderweb for hackle. Nano 24/0 as substitute for Spiderweb

Body: Fly-Rite poly dubbing no. 19

Wing: One to three strands of white poly yarn depending of the size of the fly and water you will be fishing

Thorax: One to three strands of peacock herl, depending on the hook size

Hackle: Blue dun, dark dun, light dun

Tying Steps

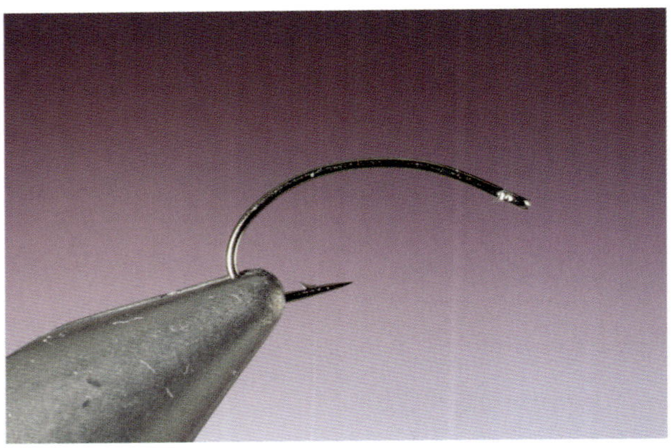

1 The new Daiichi 1160 and 1167. This curved shank, powerful wide-gape conforms to my original blueprint hook from the 1980s. Place a size 10 into your vice.

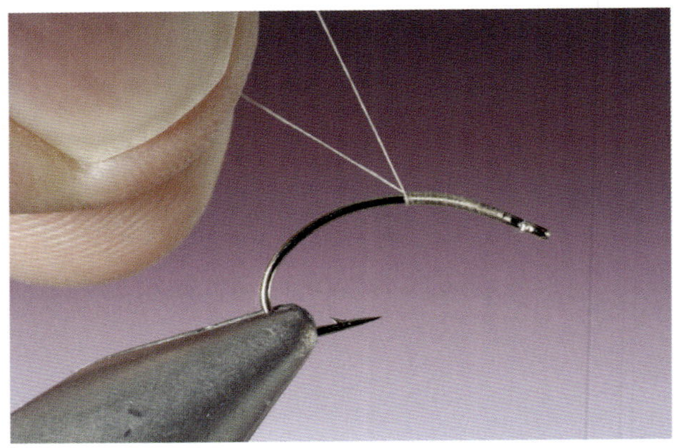

2 Wrap entire shank with thread. This foundation will help to lock the wing in place when the fly is finished.

3 Cut off two strands of poly yarn and taper one side of the yarn with your scissors before tying it in. This is to ensure that the underbody is as slim as possible.

4 Put the tapered side of the yarn on top of the hook shank as shown in the picture and secure it well at the location where the wing can pop out of the centre of thorax later.

5 Wrap your thread down to the bend and back again so you create a nice tapered underbody. Be critical of this stage! The better created the underbody, the easier to make the final body later.

6 Tie in the hackle so it lies in the same orientation as the yarn.
Tie it in with the shiny side up.

7 Clip off the protruding part of the feather shaft.

8 Form an upright wing by tying up the yarn and hackle. This must be done accurately in order to avoid potential problems with the hackle in the remaining tying steps.

9 Add some dubbing wax to your tying thread. This is because when using Fly-Rite poly dubbing, it makes it a lot easier to apply the dubbing and twist it on your thread.

10 Bring your thread to the bend and apply a tiny amount of dubbing to the thread. Use only just enough dubbing to cover the underbody.

11 Wrap the dubbing up to the base of the wing by creating a very slim but nicely tapered body. Eventual surplus you don't have to pull off but use it as a foundation for the thorax.

12 Tie in three peacock herl fibres. Tying the herls in by the tips creates a much nicer thorax. Make sure the strands are well secured in front and behind the wing. This reinforcement prevents the thorax from coming off.

13 Clip the excess lengths of the herl in front of the wing.

14 Wrap the herl, starting with three turns behind the wing. Continue the other wraps in front of the wing to complete the thorax. Secure and clip off the excess herl.

15 Make a few extra thread wraps, whip-finish and snip thread. Apply a drop of varnish to the knot.

16 Turn the hook in the vice as shown with the wing horizontal and the bend on top. Set up the Spiderweb on the base of the yarn wing and hackle.

17 Wrap a neat rigid thread base around the yarn and the hackle on the bottom of the wing. Turn the hackle so **the shiny side is at the bottom.**

18 Wrap the hackle around the base of the wing. Begin at the top of your thread base, and make each successive turn down the base. Follow Wayne Luallen's trick and let the barbs tell you when to stop.

19 Pull the hackle tip tight to the opposite direction of the wing and secure it well with the Spiderweb between the wrapped hackle and the thorax. Clip the hackle tip and tie off with your whip finisher as shown above.

20 Trim away all fibres pointing down. Take a needle and apply a drop of head cement or use a varnish applicator and press some varnish onto the knot. Let it absorb well to create more durability of the hackle.

21 Clip the wing to the required length.

22 Here is your completed Klinkhåmer Special. Marvel at the simplicity of your completed fly!

A beautiful and very rare softmouth trout that I caught with the Klinkhåmer in the Neretva River in Bosnia

Brook trout caught with the Klinkhåmer deep in the wilderness of Labrador in 2013

When grayling in Norway get over 50cm they turn into very dark colours (© Hein van Aar)

Beautiful turquoise shiny Icelandic trout caught in one of the lake feeders with a size 4 Klinkhåmer in 2015

Klinkhåmer version 2.0

When I heard that the green peacock (*Pavo muticus*) had been put on the endangered species list, I knew right away that I had to take action. I love nature and wildlife and I would never kill any living creature just to use its fur or feathers for flytying. When I hunt I do it with my camera, and my lens is my gun. Don't misunderstand me, when an animal is killed for food, I will be the first to grab its fur and feathers because I simply hate things going to waste. When I started flytying, a lot of my materials were road-kills and I know how to prepare skin and feathers well. I had a great master who taught me. I also have a lot of friends who keep birds and they keep some key feathers for me when their birds are moulting. To me, there isn't anything wrong with that! Sadly, for customs officers who follow the rules strictly to the letter at airports and borders, there is.

Is this a serious problem for us flytyers for the future? Yes it is – especially after I heard a story from a flyfisher travelling to New Zealand who had to leave all his tying materials and flies behind. More and more countries ask for documentary proof that the flies were made from materials from legal sources. So travelling with lots of flytying materials will only make things more difficult. I also know a flytyer who missed his flight because he had a simple partridge cape in his hand luggage, and security and customs didn't know what it was. He was delayed for so many hours that he missed a full day of his presentations. Although I love to tie flies from natural materials, this is the main reason I am switching more and more to synthetics. It is getting all too complicated and you never win a discussion against the bureaucrats. When I travel for international flytying demonstrations, I try to prevent problems by borrowing most natural materials from friends at my destination.

Endangered

The peacock is one of the most beautiful birds on our planet and that is what causes the problem. In the wild the green peacock is decreasing fast and it is good that it is protected on the IUCN Red List as an endangered species. Peacock feathers are still readily available from the not-at-risk blue peacock; however, we never know what the future will look like. I use a lot of peacock in my flytying, not just for the countless Klinkhåmers that I tie every year, but also for many other flies that I make as well. Many patterns within my large Bondal series of Atlantic salmon flies have a wrapped body of peacock herl, so I searched for the best substitute on the market today.

The best synthetic to replace the real peacock thorax on my Klinkhåmers and bodies of my Atlantic salmon flies, I found in using Straggle String, Straggle Legs and synthetic peacock herl, all three available from Semperfli. The Klinkhåmer version 2.0 in this chapter I will tie with Straggle String.

Tying the Klinkhåmer Special version 2.0

Dressing

Hook: Daiichi 1160 or 1167 Klinkhåmer hooks size 8-20
Thread: Semperfli 8/0 classic waxed thread in tan colour and Spiderweb for hackle or Nano 24/0 as substitute for Spiderweb
Ribbing: Semperfli 6a vintage orange pure silk thread
Body: A small bunch of natural black fibres from a large wing feather or Fly-Rite poly dubbing no. 2 (black) as substitute
Thorax: Straggle String black
Hackle: Dark or light dun cock hackle

Tying Steps version 2.0 Steps 1-6 are the same as the original Klinkhåmer (see pages 158-159)

7 Let me start with tying step 7 after you have clipped off the protruding part of the feather shaft.

8 Form an upright wing by tying up the yarn and hackle. (I am using the black feather fibres as body because so many people prefer them as body material for Klinkhåmer flies.)

9 Tie in the ribbing and natural black feather fibres. Add some varnish onto the underbody to secure the fibres and to make the herl body more durable, once it has been wrapped around the hook shank.

10 Wrap the body slim and well tapered. Begin tying deep into the bend. Wrap it along the shank and stop just behind the wing. Leave enough space for the thorax.

11 Rib body with the silk thread by making tight wraps in the opposite direction. This will make the body more durable. I usually make a little tag before starting to wrap the thread over the body. Wrap in one movement!

12 Pick your straggle string in the preferred colour. I love to experiment with the colours and Semperfli offers many choices.

13 Tie in your Straggle String before and behind the wing. I mostly take one strand and just fill up the thorax until it feels about right.

14 Wrap the Straggle String several times behind the wing, and then form a thorax by wrapping the remainder in front of the wing to create a nice looking thorax.

15 Trim away the waste and tie off with your whip finisher. Apply a drop of varnish to the knot.

16 Turn the hook in the vice, so the wing is horizontal and the hook bend uppermost. Set up the Spiderweb on the base of the yarn wing and hackle.

17 Wrap a neat rigid thread base around the yarn and the hackle on the bottom of the wing.

18 Turn the hackle so the shiny side is on the bottom. Then wrap the hackle around the base. Begin at the top of the thread base, and take each successive turn closer to the thorax.

19 Pull the hackle tip tight to the opposite direction of the wing and secure it well with the Spiderweb, between the wrapped hackle and the thorax. Cut off the hackle tip.

20 Tie off with your whip finisher around the wing between hackle and thorax.

21 Trim away all fibres pointing down. Take a varnish applicator and press some varnish onto the knot. Let it absorb well to create more durability of the hackle. Clip the wing to the required length.

22 Here is your completed Klinkhåmer Special updated to version 2.0. with synthetic thorax.

The most important tools that I use to produce my Klinkhåmers. Good quality scissors, the best dubbing wax ever made, Spiderweb, Varnish Applicator, dubbing brush and my own designed whip finisher to make it easy to tie off the hackle around the the wing case

The best of the best synthetic Peacock herl ever developed.
Semperfli made three awesome colours: black, green and blue!

Straggle String that I use a lot for making Klinkhåmers 2.0 with synthetic thorax. My favourite colours

Straggle Legs that I use for the thorax of larger Klinkhåmers

Two Klinkhåmers tied with a thorax from the brand new synthetic Peacock herl in black and blue colours

My biggest pacu in Malaysia caught on a size 2 Klinkhåmer at EAK in 2019

Catfish caught on size 2 Klinkhåmer at the Vaal river in South Africa in 2019

A big Atlantic salmon caught on a size 14 Klinkhåmer in Canada

My first ever Sultan fish caught on a size 4 Klinkhåmer at EAK in Malaysia in 2017

Repairing damaged parachute flies

I do a lot of fishing workshops with kids and I love it even more now I am retired. I use a lot of Klinkhåmer flies during my lessons because the kids can see them so well on the water. Instead of practising on a grass field first, I usually do my classes directly in the water, which works best for me. I never had a flyfishing teacher. Because the kids still have a lot to learn, they damage dozens of Klinkhåmers every year.

That set me thinking, and resulted in the idea of starting to repair them. Repairing flies from which the parachute came off or the hackle has been damaged is a very quick and easy operation. All you need to do is follow my special Spiderweb technique. With just a spool of Spiderweb, a bobbin, a pair of scissors and a hackle, I even can repair them on the riverbank. I just cut or pull off the loose or damaged hackle and hook the fly in a branch or piece of wood, with the wing in the horizontal position and bend uppermost. I am not just talking here about Klinkhåmer flies but every parachute fly that needs a proper horizontal wing. In the example below I purposely damaged the parachute hackle that I had tied for our demonstration of version 2.0. It was even harder to get it off.

Enjoy my little trick of repairing a parachute as explained in the tying steps below.

1 Always collect your damaged parachute flies. I usually wait till I have twenty or thirty before I start my repairing mission.

2 Before I store the broken flies, I carefully remove any damaged or loose hackles. The best way to do this is with a scalpel.

3 Put the hook in the vice, so the wing is horizontal and the hook bend uppermost. Set up the Spiderweb on the base of the yarn wing and hackle.

4 Wrap a neat rigid thread base around the yarn. Then tie in the new hackle with the dull side towards the wing.

5 Clip off the protruding part of the feather shaft.

6 Turn the hackle so the shiny side is on the bottom. Then wrap the hackle around the base. Begin at the top of the thread base, and take each successive turn closer to the thorax.

7 Pull the hackle tip tight to the opposite direction of the wing and secure it well with the Spiderweb between the wrapped hackle and the thorax. Cut off the hackle tip.

8 Tie off with your whip finisher around the wing between hackle and thorax.

9 Trim away all fibres pointing down. Take a varnish applicator and press some varnish into the knot. Let it absorb well to create more durability of the hackle. Clip the wing to the required length.

10 Here is your completed *repaired* Klinkhåmer Special.

Making Klinkhåmers with extra slim bodies

Ever since 1984 I have told people that one of the main secrets about my Klinkhåmer is to tie it with a slim body. When I use my regular (size 10 and bigger) tying technique, I love to use a part of the wing as underbody and use it in combination with a tiny bit of dubbing, just enough to cover the underbody. Of course I have faced moments in which fish prefer fat bodies too and that's why I have a few Klinks with thicker bodies in my fly box too. They are usually only needed in spring and early summertime.

The minority of my Klinkhåmers have a slim body and besides using poly dubbing, kapok or other extra fine synthetic materials, I also love to use a few fibres from a large wing or tail feather ribbed with pure natural silk thread. This latter combination is absolutely deadly when making extra slim Klinkhåmer bodies. Silk ribbing gives a beautiful segmentation while also making the more fragile fibres a lot more durable.

For my bigger flies tied from size 10 and up there is no need to change my ordinary tying technique when you want to tie slim bodies, but when I want to tie sizes 12 and down I prefer using another technique to secure the wing on the hook shank. In this situation I try to keep the biggest part of the hook totally clean without any underbody at all. When the wing is tied in leaving the largest part of the hook shank totally bald, you can tie extremely slim bodies. This wing technique is also very simple and here it is.

Klinkhåmer Special with extra slim body

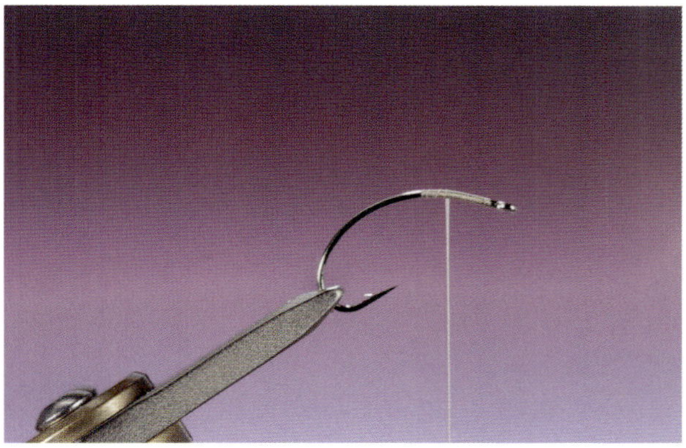

1 Wrap about a third of the shank with thread. This foundation will help to lock the wing in place when the fly is finished.

2 Cut a piece of poly yarn and take away some fibres so that the total amount will fit to the required thickness of the wing. Put it underneath the hook shank in the middle. You can secure it if you want.

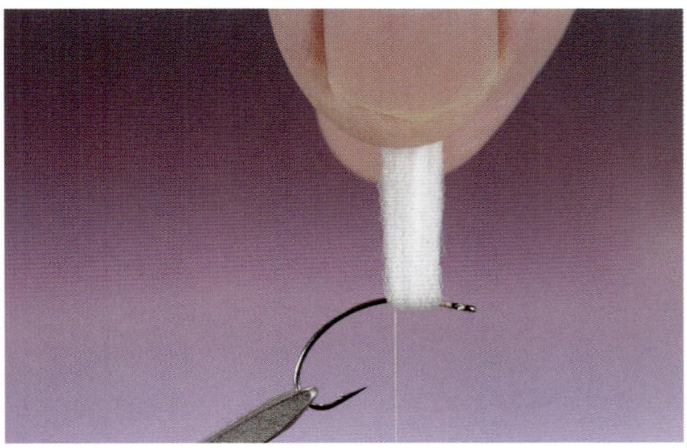

3 Pull the wing up and secure the wing to the hook shank by tightening the wing at both sides. For extra strong durability I do a few wraps in the opposite direction and around the wing-base as well.

4 If you did it correctly, it will look very similar to the picture above.

5 Tie in the hackle by securing the quill on top of the hook shank.

6 Tie up the hackle and secure it to the wing post with your thread. You now have a bare hook shank left for tying extremely slim bodies. Complete the fly same way as shown in the tying steps for the original Klinkhåmer.

Mongolian grayling readily took a size 6 Klink

What does the Klinkhåmer imitate?

I must confess that when I developed the Klinkhåmer Special, my depth and understanding of entomology was at a very early stage.

The first Klinkhåmer was meant to imitate what I found in the stomach of the grayling I took for lunch. I only knew that what I tried to imitate was a caddis pupae, right on the verge of hatching.

I first tried to imitate its curved body using a grub hook. And I could replicate the emerging legs with the parachute hackle. These elements best describe what eventually gave rise to the fly's appearance today.

Now four decades later, I view my Klinkhåmers not only as wonderfully effective flies, but as an imitation of a series of flies. I can alter the basic pattern in ways which imitate emerging sedges, midges and many species of upwing flies. In addition, my Klinkhåmers have become a perfect imitation for many different types of large, struggling terrestrials. Even when there is no visible evidence of a hatch, fish love to rise for it and take it. This fact alone makes the fly a truly great dry fly searching pattern.

Now you see it, now you don't. When the tall white wing disappears, just set the hook

Fishing the Klinkhåmer Special

When the rules permit it, I mainly fish my Klinkhåmer fly downstream. Because I learned to fish on streams in the mighty rivers of Scandinavia, I have many personal preferences and good arguments to support a downstream presentation. The simplest explanation is that I much prefer – and find it most productive – to fish a drift with the fly in front of the tippet and leader. I pretend to be able to think like a fish and thus, I believe that a lot of fish will refuse flies when they discover the leader or the shadow formed by the leader. This is very common with upstream fishing. One might argue that very fine leaders tend to eliminate this problem, but I say that very fine leaders and big fish are incompatible! I learned this very early in Scandinavia and I know from experience, that when I fish downstream, my catches improve dramatically. **I only use one backcast in my river fishing.** My catches have always been my proof of doing things right or wrong while fishing.

Because I gained so much experience in downstream fishing, I adapted my fly casting to it, and by making a combination of a parachute cast and a snake cast, I can cover extremely long drifts and multiple fishing lies. It also means that I have a lot of slack in my line through which I have to set the hook quite vigorously.

Using the Klinkhåmer as sight indicator

A lot of flyfishers use a Klinkhåmer Special as a sight indicator because of its amazing floatability. The trick is to tie an extra piece of tippet on the curved hook from the Klink where they connect a small nymph or emerger. In South Africa, I saw several anglers doing it when they were hunting for the beautiful yellowfish. In New Zealand, the trick is so popular that many call it the New Zealand rig. Also known as the Klink & Dink style.

A word about my own leaders

I gather that there are many anglers who don't like braided leaders. For the record, *I love them*! Especially the tapered hollow intermediate ones. I find they give extra power and weight to the

Huge Klinkhåmer (size 1) made for a friend to try in the Amazon. Sadly his trip was cancelled due to Covid-19

Nice brown trout seduced by a Klink in Bosnia

My exchangeable C&F fly boxes that store my long wing Klinks and Danicas perfectly without damage

leader, they resist wind knots, and they deliver the ability to cast extremely large dry flies, such as my Klinkhåmer Special, bombers, bugs, large hoppers and mice. I have used these leaders since the early 1980s with enormous success. They have saved me dozens of hours during my youth classes too because the kids struggle far less with knots than if they were using normal leaders.

My leaders and tippets are different from most systems

My technique for securing the leader to the fly line is to push my hollow braided tapered leader about 5cm over the fly line and tie the end off with tying thread. For the record, I have *no* loop-to-loop connections, because I saw too many of these connections fail when people were playing really big fish. Loops are strong at first but with time during fishing becoming weaker. I just add a drop of waterproof superglue to protect my tying threads and it never comes off again. All my leaders are tapered, going down from 50-20lb for big fish and from 30-10lb for normal fishing. I mostly have a strong silver or stainless steel tippet ring at the tip end and all individual leader parts are connected without using knots. I put them together with a special needle. On the tippet ring, I attach one size tippet material only. I have found these leaders extremely handy, smooth, durable, and unbelievably strong in their connection.

I use exactly the same system (although thicker in diameter) for catching tarpon and sailfish and therefore even dare to call these leaders tarpon- and sailfish-proof. These extremely powerful leaders no longer have a tippet ring and their leader and tippet are connected in a different way too, because you often need shock tippets or a bite tippet in very different diameters.

In all my fishing catching some very large fish, I have never had this leader connection snap and I have used it intensively since 1981. I set up two of these leader systems in Malaysia for two guests on the same boat and they caught 29 and 32 sailfish in one week without any breakages. When something breaks, it's only the tippet or the fly you lose. I do a lot of youth classes and kids break or tangle their tippets a lot but with this system I replace a new tippet in less than a minute. If you use a parachute or snake cast, or a combination of both within the same cast, you will be fishing the Klinkhåmer in almost the same way as I do! I often pull back the line a bit before it has landed to create even more slack in the line when fishing downstream in big rivers.

Klinkhåmer catches so far

Coarse fish (Europe)

Roach
Rudd
Dace
Ide/Orfe
Chub
Common Carp
Grass Carp
Leather Carp
Crucian Carp
Mirror Carp
Barbel (Spain only)
Bass (Portugal only)
Hybrids (Netherlands)
Pumpkinseed (*Lepomis gibbosus*)
Common bleak (*alburnus-alburnus*)
Asp (small size only)

Exotic species

Catfish (South Africa only)
Pacu (Malaysia)
Sultan Fish (Malaysia)
Jungle Perch (Malaysia)
Sikang (Malaysia)
Tarpon (in river) (Taiwan)
Cichlids (Malaysia)
Yellowfish (South Africa)

Salmonides

Grayling (Europe, Canada, Asia (Mongolia))
Arctic Grayling (Northern Scandinavia, Canada, Alaska)
Atlantic Salmon (Norway, East Canada)
Seatrout in saltwater (Denmark and Norway)
Seatrout in river (Europe)
River White fish (Coregonus, Scandinavia)
River White fish (Canada, alot different from Europe)
Lake White fish (Scandinavia and Canada)
Brook trout (Canada)
Char (Scandinavia, Iceland and Finland)
Arctic char or searun char (northern Scandinavia)
Rainbow trout wild (few rivers Europe, Canada-Alaska)
Rainbow stocked (ponds)
Tiger trout (pond only)
Lake trout (Yukon and Labrador)
Dolly Varden (BC Canada and Yukon)
Softmouth trout (Bosnia only)
Amur trout (*Brachymystax savinovi)* (Northern Mongolia)
Blunt snouted Lenok (Mongolia)
Sharp snouted Lenok (Mongolia)
Grayling (Mongolia different from Europe, Alaska and Canada)

Casting luck

Toman or Gigantic Snakehead (Malaysia) casting a Klink into his mouth when he came up to breathe and I hooked and landed him! Four witnesses, hahaha.

Klinkhåmer with orange wing for difficult sight conditions

Klinkhåmer version 1989

Klinkhåmer version 3.0 tied in 2023

Three PPSs taken straight from my fly box

CHAPTER 7

Para Poly Sedge

Since I retired, I have plenty of time to go through all my fishing and flytying diaries and it is quite amazing to see how productive I was in the 1980s and 1990s. Some claim that the 1980s and early 1990s were the golden years in flytying because everything seemed possible in those days. Without the fast and incredible world of Internet we had to rely on books and magazines to gather information. A lot has changed since then, but personally I think that every flyfisher and flytyer goes through different stages anyway, and in my life I was most creative and went through my most productive flytying stages between 1980 and 1999. I tied and developed most of my patterns in those years, but my ideas and improvements have never stopped. I just do things more slowly now.

Early notes

Once I opened one of my old diaries and read all my early notes about a sedge pattern I had named 'The Para Poly Sedge' (PPS). It is another parachute fly that I developed just a few weeks after I tied up my first Klinkhåmer Special. But it was a totally different pattern because it was a Parachute Sedge. While reading these old notes I re-lived the magic and excitement of the pattern and it gave me the idea to write this chapter. It's always interesting to learn the history of a particular fly and how it was developed.

The first examples of the PPS were the result of some tying experiments during a terribly dark, cold and rainy summer that was blighting the eastern part of Norway. These were days when you didn't want to be outside. The temperature had dropped dramatically and a raw eastern wind, hail and wet snowstorms didn't spell much good. Although we normally camp out beside the riverbank we had decided to rent a comfortable cabin. The bad weather had already spoiled our wilderness camping and almost all our clothing was wet, a situation where a warm and luxurious cabin is the only solution.

I had just put some wood on the fire and my wife was relaxed, reading a book beside the open fireplace. The fishing was extremely poor so I concentrated on flytying by creating a special sedge pattern for a small tributary of the mighty Glomma river that I had never fished before. A good fishing friend, who had died that winter, had fished this tributary frequently and we had just made serious plans to fish it together during that summer.

Life's lessons have taught me that you should never ignore a good tip, especially when it comes from one of your best fishing friends. I knew the river was famous for its excellent caddis hatches so I tried to develop a universal sedge pattern. Besides caddis, the river was also teaming with Baetis larvae and I also love to tie realistic *Baetis rhodani* lookalike nymphs, and given the weather conditions there certainly was enough time to play around with thread, fur and feathers.

New Sedge pattern

When I looked through my spare box containing various prototypes of my Klinkhåmer Special that didn't make the mark, I suddenly saw a great opportunity to create a nice Parachute Sedge pattern using the same tying technique. I tied up a few new variations and after several attempts, and few hours later, I felt pretty happy with the results and was eager to try out my new experimental fly.

The Glomma River in those days could easily handle a lot of rain without colouring and its water level rises very slowly. Small tributaries, however, are a different matter and their watercourses can change quickly, especially after days of rain. The tributary I wanted to fish still looked okay that morning, but unfortunately it was no longer fishable in the evening. Thunderstorms in the highlands had changed this peaceful little stream into a wild, turbulent river.

All that I could do was go to the main river to check my prototype patterns for their floatability in rugged water and test some retrieving techniques in some lakes nearby. From the lake tests, I learned the pattern worked extremely well in imitating caddis flies walking on the surface. The rain continued and so I tied a whole bunch of my new Para Poly Sedges and ended up filling up my fly boxes.

The truth about the success of my PPS fly is unusual because I didn't get the chance to test it in any objective way while fishing this particular river. The following days there was still too much water and fishing was hardly possible. Wading was too difficult – even dangerous – and the best pools were completely flooded. Access to the river's hot spots was simply not feasible and the fish were impossible to spot. These were probably the worst fishing conditions I had ever encountered in this place – and remained so for many days.

Enthusiastic response

Of course, I still tried some fishing, but not surprisingly, without success. The fish were not interested in my new sedge imitation or any other flies. My experimental flies were eventually simply forgotten among my other patterns. Every reasonable flyfisher agrees that one cannot condemn a river as poor just because an experiment failed due to the weather. I proved that right, because a few years later I did

Beautiful trout caught in the Neretva river in Bosnia

PPS test in Mongolia was an enormous success

extremely well on this particular river in better weather conditions. Meanwhile the PPS stayed in my fly box without being used very much. To be honest, I only used it maybe a dozen times.

During trips and in conversations with other anglers, I gave away quite a few flies and several of these were PPSs, simply because I had so many of them. Two years after I first tied them, my attention was drawn to the pattern in the following way; I received an enthusiastic letter from a German flyfisher to whom I had given a few PPSs to try in Scandinavian waters. He and his friends had done extremely well with the fly and he claimed that he enjoyed his best ever trip to Norway and Sweden largely thanks to this pattern. The PPS had become his absolute favourite and today he keeps a fly box full of them in many colour variations. The same year, another letter reached me with favourable reports on the pattern, this time from the UK. It seemed that other anglers were having much more success with this pattern than I had. However, after that second letter, I decided to give the fly closer attention and tried it out more intensively in various parts of Europe. This time my luck had changed and the results were far better. Most spectacular of all was catching a few grilse in a small Norwegian river. I even had some really good success catching seatrout while fishing in a few estuaries and sea pools of the Fosen Peninsula in Norway. I also fished the PPS for many years in Canada and it became my ultimate sedge pattern, especially for stillwaters.

How to fish the PPS

The fly must be fished in the surface film, with the polypropylene yarn wing above the surface. Once the wing is treated with Fly-Rite's dilly wax, the fly floats very well and can be fished in drag-free drifts, or while using upstream retrieve techniques. The parachute hackle up front gives added attraction to this pattern and leaves a nice wake if the fly is retrieved at the right speed in slow running or stillwater. Using slow retrieve techniques made this fly an ultimate killer in the lakes of Scandinavia and Canada.

Tying the original Para Poly Sedge

Dressing

Hook: Daiichi 1280, 1260, in sizes 8-10-12-14

Thread: Semperfli 8/0 classic waxed thread in tan colour.

Body: Fly-Rite's poly dubbing in the following colours: light tan, dark tan, rusty olive, medium brown, dark brown, golden olive and any other colour to match the hatch.

Wing: Several strands of Semperfli's poly yarn in white, blue dun, light dun, cream, brown or mixed

Hackle: Good quality light ginger, ginger, light brown or blue dun

Wingcase: Poly yarn same as wing

Note: For the wings and wing case, I also prefer certain blended colours of Predator fibre because it works exactly the same as poly yarn when treated with Dilly wax. It's ultimately the same material and you can mix it a lot easier than poly yarn strands. I love blended colours for example Black Nose Dace Semperfli Predator fibres as I used for the flytying steps of this PPS. It's good to show you that you can achieve exactly the same with Predator fibres too.

Tying Steps

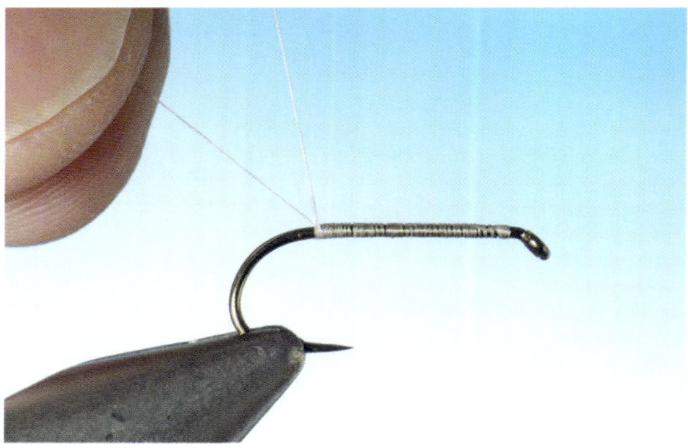

1 Place the hook in the vice and wrap the entire shank with tying thread.

2 Cut a piece Predator fibres of approximately 5-6cm long and taper one side at an angle of 45 degrees.

3 Put the tapered side on the hook shank as shown in the picture above.

4 Secure the fibres well with your tying thread all the way up as close as possible to the hook eye and back again until you have created a nice tapered underbody.

5 Tie in a good quality hackle pointing in the same direction as the wing (i.e. with the feather extending forwards) and take the thread back to the hook bend.

6 Put some dubbing wax on your thread and set up your thread with Fly-Rite poly dubbing in the colour of your choice and dub a nicely tapered body. Wrap your thread back to the position as shown.

7 Grasp the tuft of Predator fibres and wrap the hackle around its base.

8 Take each successive turn closer to the hook shank. This ensures a compact, well-compressed hackle.

9 Pull the hackle tip backwards over the top of the hook shank and secure it with 2-3 windings around the body about 6mm behind the eye of the hook. *(See red circle.)* Trim away the excess hackle tip, but do *not* cut your thread.

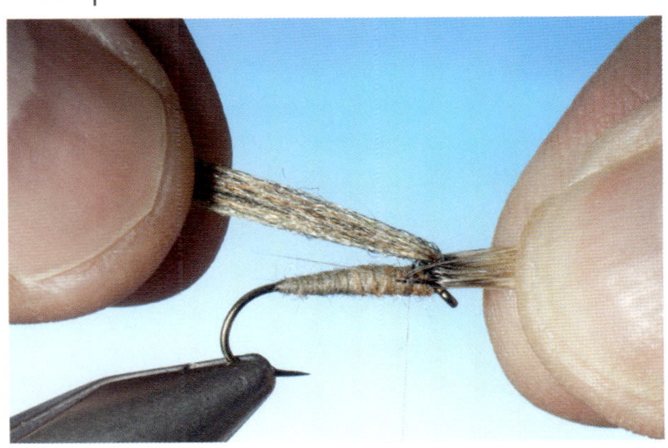

10 Now, carefully groom all the hackle fibres forwards, then pull the Predator fibres backwards between them.

11 Tie down the wing where you secured the hackle and lock the wing with your tying thread.

12 Tie off around the wing and body with your whip finish tool.

13 Trim the poly yarn wing to a length that is about 1-1.5cm longer than the bend of the hook. (Experiment with the length of the wing during fishing.)

14 Apply some lacquer by using a varnish applicator on the windings and your fly is ready!

The River Soča in Slovenia where the PPS can be unbelievably successful at times

CHAPTER 8

The Culard and Once and Away

Cul de Canard against all rules

Cul de Canard (CDC) feathers have risen to worldwide popularity over the last five decades. To avoid confusion about their origin I can tell you that the idea to use these feathers for flytying actually came from Central Europe. Professional fly dresser Henri Bresson from France invented the term 'Cul de Canard' for a fly, which he created entirely from CDC feathers. This was somewhere in the late 1950s. CDC has been used by anglers in the Swiss and French Jura for almost a century. The late Marjan Fratnik from Slovenia popularised CDC in the early 1980s with publications on his F Fly series. Marjan became inspired after he read Jules Rindlisbacher's book, *Der Practische Fliegenfisher* (1970). It is one of my favourite books and Jules was probably among the first authors who wrote about CDC in a language I understood.

Another great flytyer who is world famous for his superb CDC designs is Marc Petitjean. He kept the Swiss tradition foremost and designed new patterns continuously. Being a very close friend of Leon Links has a lot of benefits and is extremely handy when I needed to know something about CDC. For me Leon is one of the top experts on CDC in the Netherlands.

Leon's book on CDC flies

My CDC background is very limited, but if you want to know the history of CDC you should read Leon Links' book *Tying Flies with CDC – The Fisherman's Miracle Feather*. It is a marvellous book on CDC, published by Merlin Unwin Books in England and available as an e-book. Over the last forty years CDC has had a lot of attention, especially in the UK and the USA and suddenly there seem to be a lot of flytyers who claim to have been the inventors of the CDC flies.

My way with CDC

The Dutch have been experimenting with dressings incorporating CDC especially since fishing journalist Kees Ketting popularised CDC in the early 1980s, developing and refining patterns for all types of fishing.

I have used CDC for at least forty years and have tied many different patterns with it through all these years, but I mainly fish with just a few patterns. In Leon's book, you will find some extra rugged CDC patterns that I use in the wild rivers in Scandinavia. To be honest, my use of CDC flies seems to go in cycles; there are periods, or even seasons, when I hardly fish with them, and there are times when I use them intensely. It depends a lot on where I am fishing. In warmer countries like Italy, Slovenia or the Balkans, I use a lot more CDC flies then when I am fishing in Arctic or subarctic waters. Of course, the old and original CDC patterns are excellent flies. Nevertheless, I prefer to use my own patterns. I also like to experiment and try to improve patterns. I believe it would be very hard to design a completely new

The rivers in Slovenia are perfect for trying flies like the Culard and Once and Away

CDC pattern, but some improvements, or design differences, are always worthwhile describing.

My different style of fishing and my flyfishing thoughts and strategies need special patterns. The fact that my distance vision isn't that good and my preference for powerful rivers and always broken surface water is reflected in most of my dry fly designs. Over the years I have also found some very good uses for CDC in lakes and very quiet waters. Today I never stroll along a river, lake or reservoir without CDC flies in my fly box, even though I may only have a few different patterns with me.

Floatability

In the past the greatest problem with CDC was that many people didn't know where CDC feathers came from and where they should be plucked. Today people are much better informed. CDC feathers are found near the preen gland of water birds. The CDC from wild ducks appears to be the most popular and best. The colour is absolutely incredible and perfect for matching a multitude of hatches. The fresh feathers of wild ducks are light and naturally coated with the oil from the duck's preen gland, and the natural oil makes those feathers extremely water-repellent **and their lightness and buoyancy makes them invaluable for developing dry flies and special emergers. However,** further investigations prove that the *structure* of the feather actually has a lot to do with the unique and extremely high buoyancy of this very special feather. A combination of the natural oil and feather structure make CDC so attractive to the flytyer.

When I received a very large amount of white CDC from Peking ducks many years ago, I started to experiment with dyeing and I never had any buoyancy problems when I used dyed feathers, which reinforced my confidence in the feather structure theory, because dyeing CDC means that all the natural oil will disappear. Under a microscope you can easily see that each feather has a great number of barbules. Those multiple barbules aid buoyancy by greatly increasing the surface area of the feather and it makes them much stronger and more durable than most of us believe. I often compare CDC fibres with spider web thread that I use for my parachute technique. Wrapping once is very fragile, but durability improves greatly with added numbers of wraps. I do the same by adding extra CDC feathers to my flies.

Mobility

The fine fibres of a CDC feather are also probably the most mobile you will ever encounter, which is why I use them a lot in my nymphs and wet flies. In the slightest breeze they also flutter to simulate life and for this reason they make a perfect wing imitation. The wind also has a great impact on the mobility of the floating fly. In or just under the surface the fibres improve the mobility of the pattern and give an excellent lifelike appeal. Size, shape (silhouette) and colour are the most important facets of imitating insect life, but when I started to use CDC I added *mobility* too. This is why I never cut the single excess barbules away from my flies. CDC is ideal for both emergers and dry flies. Although the original CDC patterns are good floaters, if they are tied well, they are not unsinkable as is so often claimed. Several of these patterns are tied very sparsely and easily become waterlogged after catching a fish. If you don't treat the CDC fly well after hooking a fish, it will sink easily, especially in rapids and turbulent currents. Amadou helps you a lot to dry the fibres and I love it, using exclusively Revolution Amadou from Romania. Floatants will stick the fine fibres together and will destroy the effectiveness of the pattern because they then lose the barbules effect. Therefore I never use any (sticky) floatant on my CDC patterns.

I used some CDC oil in the past, but I would point out that with normal or sparsely tied CDC flies it is often much easier to tie on a fresh fly than to struggle in order to dry the one in use. This has finally led me to use CDC against all rules!

Before I write about my own CDC-patterns, which are certainly not lightly dressed ones

The outflow from the a big lake in eastern Norway, a wonderful spot to experiment with CDC flies

The best of the best: Revolution-Amadou Premium (front) and GTX made in Romania

– you should not underestimate the value of the extremely lightly dressed CDC patterns developed by today's experts or tyers from the past. In certain waters these patterns can be extremely successful at times, rather akin to a good Yorkshire Spider that loses its main strength when it is overdressed.

With many CDC patterns it is not different. During the years I accidentally discovered that some of the old traditional CDC patterns are actually fished better just under the surface film and today I still use some of them in this 'emerger' role, in preference to fishing them floating on the surface – particularly with lake fishing.

My first serious 'Cul de Canard' experience

While river fishing with small, lightly dressed CDC flies I often had difficulty seeing them in faster and more turbulent water. As a result, I tended to use them in rivers with slow currents or on lakes. The first real breakthrough with my own CDC flies happened when I tied a Rackelhanen lookalike mainly from CDC. The natural colour of the CDC was perfect for imitating dark grey sedges. I named the pattern the Rugged Caddis and it was built up in four sections, each including four CDC feathers. It is a very expensive fly because good quality CDC isn't cheap nowadays. The pattern originally had a big, long wing of at least 16 CDC feathers, mainly because I assumed that the more CDC I used, the better the pattern would float, and it did for a little while. To create the sedge shape I simply cut the feathers to form the particular sedge wing I was imitating. Strange as it may sound now, I was often using the Rugged Caddis to catch Atlantic salmon in the northern part of Norway. I even tied this pattern on a small fine wire streamer hook, and at that time it was rather effective. As I began to experiment with smaller sizes and reduced the length of the wing, I used less CDC.

Then I discovered something surprising. As I shortened the wing, I got better catching results. In the end I was using a pattern that had a wing of only half the length of the hook shank, and it was still floating extremely well. I used this idea with small flies and therefore the fly was difficult to see on the surface, but because of the aggressive takes it produced, I didn't miss too many fish. When I finally started to use even smaller hooks,

I added a hackle – and the Culard Sedge was born! Today I call it simply 'The Culard' and it is the smallest fly in my entire fly collection that I developed myself. I use the Culard the most in Holland, Germany, Italy, Slovenia, the Balkan countries and high summer in Scandinavia.

CDC against all the rules

My tying technique is definitely against all the rules. I keep the stiff quill(s) of the CDC feather in the middle and do not take only the soft fibre tips as many tyers recommend. I agree with them, but I have my own ideas too. Many people said that CDC feathers should never be cut, but I still do and must confess that I prefer this method for my particular style of flies. After tying in the feathers with the quills still inside, I trim their tips. However rebellious, this is the technique that suits me best. **It is the greatest secret behind many of my CDC patterns**. Using this technique, the stiffer wing dries and 'fluffs up' with only one or two false casts and that's my biggest reason for using this unusual cutting technique. With my Once and Away patterns it even works better. This is not easily achieved when only the finer fibres are used. The filament feather tips tend to stick together and are almost impossible to dry by false casting. In my patterns the quill provides this added stiffness. The resultant Culard has become one of the best and preferred flies among my small dry fly patterns today. The pattern is not only an excellent fly in the rivers in central and southern Europe, but also works very well during high summer or low water conditions high up in the north. On lakes the fly works perfectly while floating on the surface, but sometimes it is even more deadly when I pull it through the surface or I moisten it and fish it as a tiny emerger just under the surface.

During one of my flytying demonstrations I discovered another secret of the Culard by accident. It happened when I dropped one of the ready-made Culards in a glass of water to show the public how it floats. Under the wing a bubble of air conveniently produces a fine halo-effect, which could be another reason for the unbelievable success of this tiny and easy-to-tie CDC pattern.

The Culard is the smallest pattern that I ever developed

Tying the Culard

Dressing

Hook:	Daiichi 1180, 1182 or 1190 barbless size 18-24
Thread:	Semperfli Classic waxed thread, black
Body:	Two or three fibres from a black peacock wing feather or substitute
Rib:	Semperfli yellow or vintage orange pure silk thread
Wing:	CDC feather pulled together and cut halfway up the body length
Hackle:	Dark blue dun, very fine and much smaller than usual, or starling body feather for the patterns that I want to fish just below the surface film

Tying steps

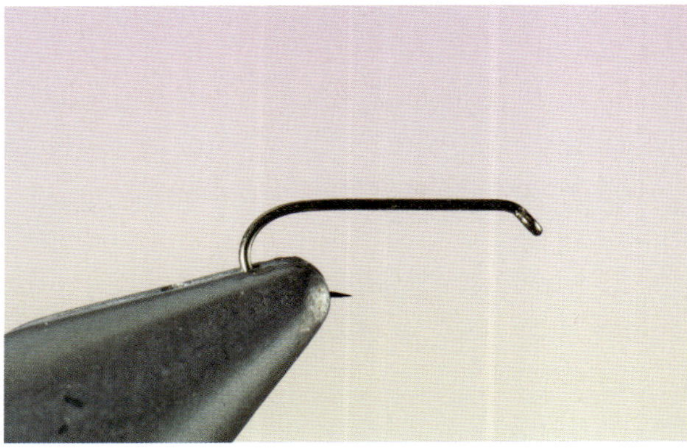

1 Pinch down the barb of the hook if you wish and insert the hook carefully into the vice.

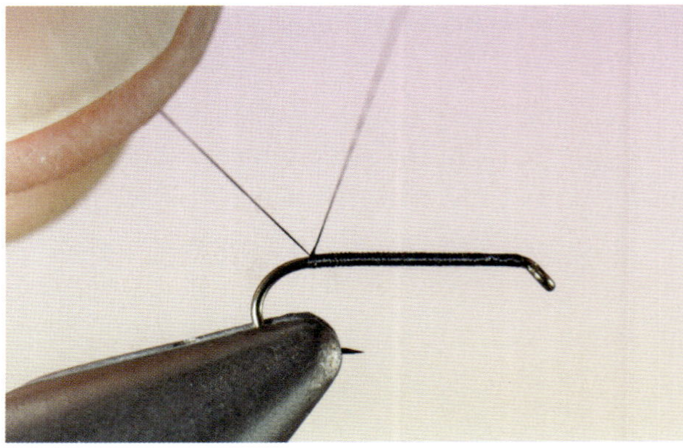

2 Wrap the shank with your tying thread.

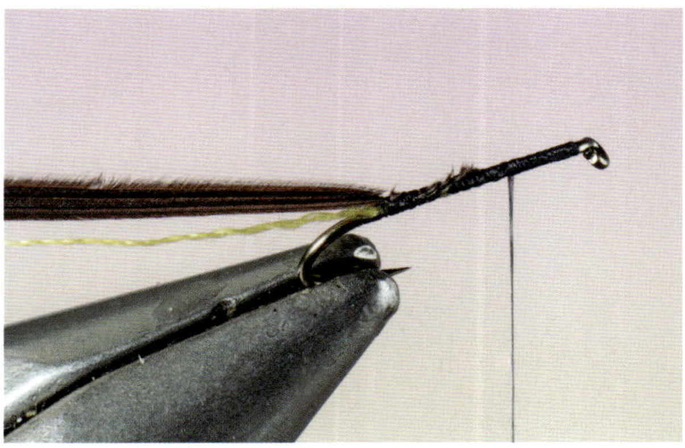

3 Secure a piece of yellow silk thread for ribbing and tie in some fibres from a black peacock wing feather.

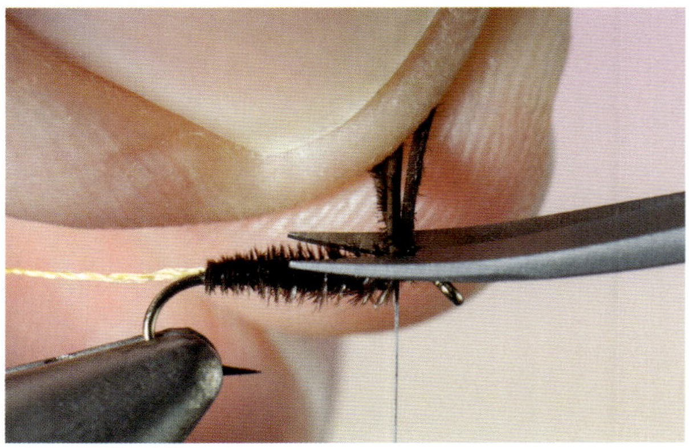

4 Put some varnish or wing cement on the shank, wind the herl fibres towards the eye and secure them with your tying thread. Be sure you leave some space for the hackle and wing. Snip off the waste fibres.

5 Rib the body with the yellow silk by wrapping it in the opposite direction towards the eye of the hook.

6 Tie in your CDC feather (or feathers) like a sedge wing.

7 Tie a small, dark blue dun hackle on top of the hook shank.

8 Give only two or max three wraps of hackle and secure the hackle with some thread windings.

9 Whip finish and cut the thread.

10 Cut off the CDC wing exactly in the middle of the body. I also cut the wing at the riverside after first trying out the pattern with the long wing.

Dying CDC is a lot of work but the final result can be awesome

Success while fishing a strong wind-lane in Yukon, Canada

The Once and Away designed for slow running waters and lakes

Fishing a nice wind-lane during midsummer night is absolutely amazing

The rise of the Once and Away

The next CDC pattern I want to talk about is the 'Once and Away'. This fly I find one of the best and most beautiful emerger patterns I ever made and it is specially designed to be almost unsinkable. Therefore it is suitable to all kinds of water. Unfortunately most people tie the fly wrongly, which is why it needs some extra attention.

I developed the Once and Away in the early 1990s during a few days fishing in a private rainbow-stocked reservoir in Holland. In those days, I organised a good number of flyfishing workshops and a pond with fish was helpful to reach my goals. The fish were fairly educated as they had been in the water for several months and the pond was heavily fished by a great number of very experienced reservoir flyfishers.

On the first day of my trip, there was a lot of wind and fish were hard to catch. After a few hours without any success, I decided to try the opposite side in front of some trees where I could stay out of that nasty wind that was torturing my face continuously. Here I saw fish jumping and even some showing in a head-and-tail rise pattern, especially where the wind hit the water. Immediately, I started to fish with my Klinkhåmer, which is usually a deadly fly for head and tailing fish. I had some responses and missed one fish when I tried to play with the fly in the surface.

I was convinced that these fish were rising to small flies that were being blown from the trees onto the water. After another half hour with no fish, and just when I started to doubt I would get any takes at all, a big rainbow came up to the surface just a few feet in front of me. He took something, hit the surface and disappeared into the depths again. There was only one explanation for this behaviour; this fish was feeding just under the surface and NOT interested in my Klinkhåmer at all. I changed the fly to the Culard, fished it as an emerger by pulling it purposely below the surface and managed to catch nine fish that afternoon.

I was still not satisfied because too many fish refused to take the fly and turned their head when they came close to it. So the next day I caught a few of these remarkable emerging insects. Their bodies were thin but the thorax seemed big and larger than usual. Returning to my car, I tried to tie a suitable imitation. This was not easy as I had very few things with me. I had a small bobbin with threat to secure my braided leader on my fly line but I didn't have any hackle material and hooks so I cleaned a few old midges that were tied on size 14 and 16 Klinkhåmer hooks. My plan was to create a fly with a body from tying thread and some CDC feathers that I had plucked off a dead duck I found beside the road that morning.

After a few attempts I eventually managed to put something together that I thought might be successful. It was with considerable interest that I tried it. My confidence in it was established within the first few casts. In just one hour I caught more fish than in the whole of the previous day. I called the fly the Once and Away, since I had a great deal of difficulty in getting the pattern to float again after it had been dragged down by a fish. My only option left was to tie a few more while sitting in the back of my car and playing with just a bobbin and thread, flies I took to pieces to get my hooks, and a bunch of CDC feathers. The good thing was that the fly was easy and quick to tie.

Improving the Once and Away

Later when I returned home I changed the dressing to a better-looking and more durable pattern. I added a thorax from peacock herl which gave me even more faith in the new fly. To find a reasonable solution to get a better floating fly after it had caught fish was not at all easy and drove me almost crazy. The answer came from my Rugged Caddis and Culard. Adding a few

A lake trout with damaged mouth grabs the Once and Away in dead calm water in the Yukon Territory in Canada

more CDC feathers to make the wing far more bulky, and a simple cutting operation on the fly's wing design took me weeks to find out. I didn't cut the wing into any shape. I just shortened the wing and again I purposely left the stiff quills inside. While cutting more and more I saw how the action with my scissors was pushing the fine barbules into the perfect silhouette and shape that I actually wanted.

After each cutting operation I dropped the fly in a dish of water to see the results. The new shape became my best alternative for parachute-style flies but tied without any hackle. Once more I had developed a CDC pattern where CDC was used against all the rules.

The body solution

I keep the body as slim as possible, at first by only using tying thread and I rib it with pure silk. I also tried a long moose hair or quills for a little time with good results. However, when a good friend from the US sent me some bristly peccary (also called javalin or javelina) skin with amazing hairs on it I tried it and then the body was not only extremely strong it was unbelievably beautiful too. It was surely another of these Eureka moments in my life! The gorgeous natural blending colouring of cream and brown surely attracted more fishermen than fish and in that moment the Once and Away became the most beautiful emerger I ever made. The hair was extremely easy to wrap and the segmentation was simply awesome. Nowadays, you see a lot bodies made from peccary. A lot of emerger bodies do extremely well with peccary as the body, including my Klinkhåmer Special.

Making the wing

I take several CDC feathers for the wing and wing case and leave the stiff quills inside when I cut

Peccary hair was another magic touch and very strong and extremely durable body material

them after finishing the tying process. With this technique the fly is not only very easy to dry by a few false casts, but it also keeps floating really well. The visibility is unbelievable and, because I use my own curved hooks, the hooking power is superb. There were no casting problems and a perfect landing ensures that this pattern always hangs perfectly in the surface film.

The other benefit of this pattern is that it has a similar effect to my Klinkhåmer but fishes a lot better in stillwater. The wing mostly has a good bubble of air around which makes it a perfect emerger. The fly also works extremely well when fish are feeding for midges.

Using the wind

This fly will even do a lot better with precise casting and making use of the wind. With a little practice you can play around with letting the fly dance with the wind while twitching the fly with little rod tip movements. Let the wind impart some extra action and movement to the pattern and I am sure you will be amazed. I prefer to fish it in or very close to a wind-lane. The wind automatically brings it to the feeding fish. Today I use the Once and Away in rivers and in lakes all over the world. Even in rivers with a turbulent surface movement, it is possible to use the fly.

Tying the Once and Away

Dressing

Hook:	Daiichi 1160 and 1167 Klinkhåmer hooks size 10-20
Thread:	Semperfli Nano silk 20D white or black
Body:	One hair of a peccary or javelina
Thorax:	Up to three strands of peacock herl depending hook size
Wing:	Four large Cul de Canard feathers
Wing case:	Same Cul de Canard feathers as wing

Note: I prefer strong Nano for this style of flytying

Tying steps

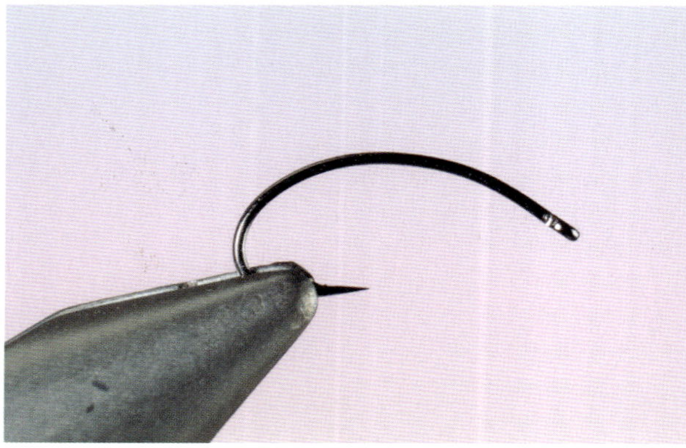

1 Pinch down the barb of the hook if you wish and fix the hook in the vice.

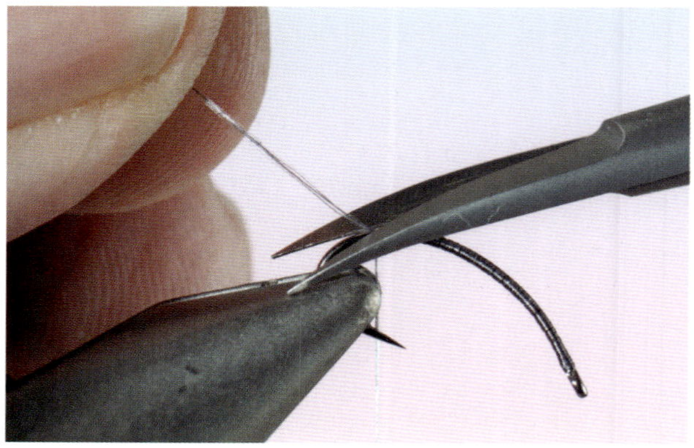

2 Wrap the shank with the Nano and cut excess thread. This avoids the difficulty of a slipping wing when the fly is finished.

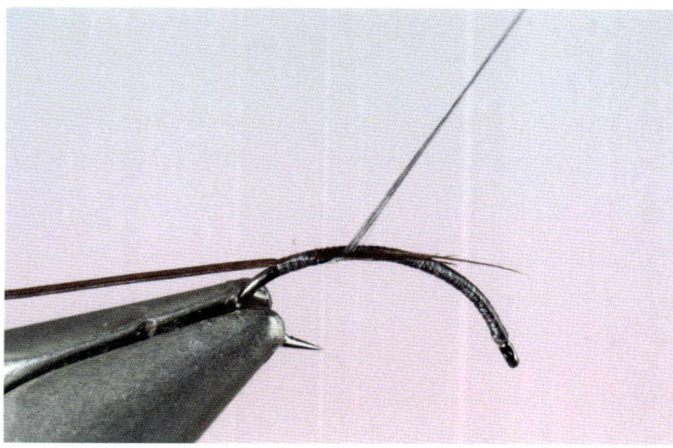

3 Tie in the peccary or javelina hair at its tip. If you can't find any peccary, use synthetic peccary quills from Semperfli. I keep the tips from the peccary for mayfly tails.

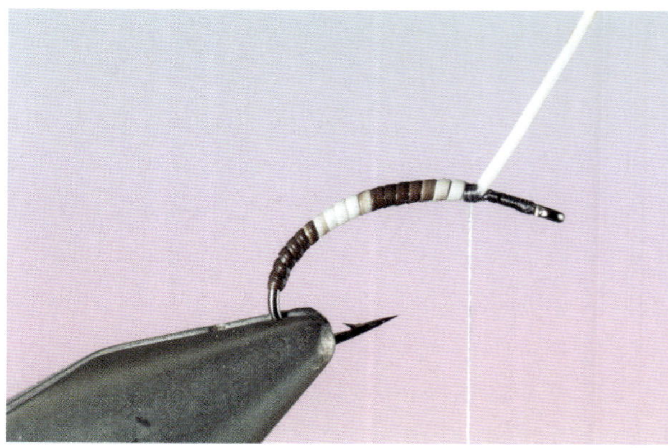

4 Wrap a nice segmented body. Leave enough space (1/3) for the next tying steps.

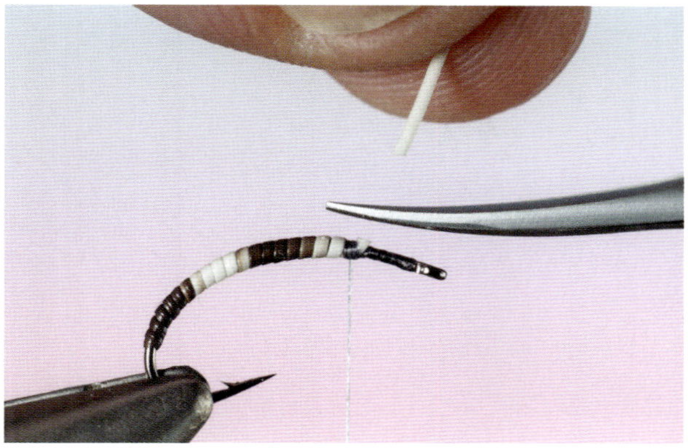

5 Snip off the excess peccary hair.

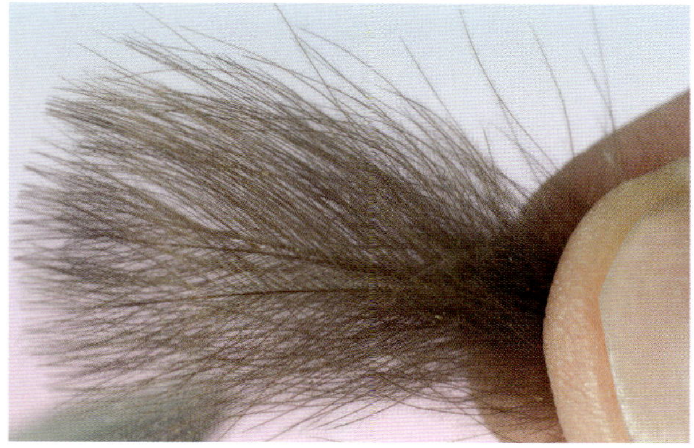

6 Select four large CDC feathers and align the tips.

7 Tie in the four large CDC feathers as shown in the picture.

8 Clip the excess CDC and tie in three peacock herls for the thorax.

9 Snip off the surplus peacock.

10 Wrap the peacock herl into a nice looking thorax.

11 Cut off the excess peacock.

12 Pull the CDC feathers over the thorax and tie them down just behind the eye with maximum four wraps of tying thread.

13 Stroke the feathers into an upright position and keep them there by applying so much tying thread in front of the wing that the wing stays up all by itself.

14 Check the position of the wing. If necessary, pull the wing feathers to the rear again and add more thread.

15 Tie off with your whip finisher and trim the feathers into a tuft. I like a wing tuft about one and 1½-2cm long.

16 The completed fly.

Note: Never cut the fibres that are sticking out of the wing case because they hold bubbles of air and make the fly more productive.

Perfect conditions for the Once and Away in early spring in Finland

An Ugly Bug and a simple Parachute Ant both tied with a sight indicator

Chapter 9

The Power of Ants

Many fishermen worldwide know that ant imitations are deadly fly patterns for trout, grayling and dozens of other species as well. Not so long ago I even caught a huge carp on a tiny parachute ant while showing a non-fisherman a few casts to demonstrate to him what flyfishing is like in the pond in front of my house. A lot people believe that you can only use ants during the summer, or at times when they are actively migrating. Maybe that's why the flying ant is probably the most popular of all ant imitations. Of course, all things considered, it sounds reasonable because during their migrations trillions of flying ants land in rivers and lakes.

Ants are excellent food for fish and many ants are eaten before they drown. To tell you the truth, I don't have a particular season for using my ant imitations, and I certainly don't use them on salmonids only. For me, an ant pattern is simply one of the most effective extra flies that I keep in my fly box today. I have dozens of them in many different colours, sizes, and shapes and I just pick a few from my large C&F collection box and put them in my ordinary fly box before I go out for fishing.

Early experiments

The first time I fished ants was when I was around 10 years old and fished a lot with realistic insects as a float fisher in the south of the Netherlands. Besides natural caddis, bees, damsel and dragonfly larvae pinched on a hook, I used living ants as a bait or trigger fish by throwing them on the water to provoke aggressive surface feeding coarse fish like rudd (*Scardinius erythrophthalmus*), ide (*Leuciscus idus*), dace (*Leuciscus leuciscus*) and chub (*Leuciscus cephalus*). These were the most active surface-feeding fishes in our local rivers in those days. I learned the trick from a retired man who I loved to fish with as a kid. Each time we threw in a dozen living ants, the fish went mad and started to feed like crazy.

Later between the ages of 12 and 15, and being a fanatical float fisherman with more creativity, I glued tiny living ants onto my hooks using synthetic bait called 'Rubiver Dur'. It was the number one bait for river float fishermen in the 1960s as far I can remember, at least in the rivers where I grew up. So, yes, I was quite accustomed to using realistic insects at a very early age and it certainly set a sound basis for my thinking as flyfisher many years later.

In flyfishing I used my first ant imitation during a visit to the River Klara in Sweden in the early 1980s, but since then a lot has changed. My ideas about ants, and therefore my imitations of them as artificial flies, went through several evolutionary changes during the 1980s and 1990s.

When a famous author asked me a few years ago which dry fly I used most of all in the Netherlands, I completely amazed him when I answered, 'The Parachute Ant'. At first he couldn't believe it! He thought I was kidding, but after I told him my story, he just shook his head. Since then, we have fished many times together, and after he saw my success with them, the ant pattern is now among his favourite dry flies too. Maybe I can convince you too about the unbelievable power of the ants.

Investigations and observations

In my flytying room, I have observed dozens of real ants. I am not really proud of that, because I respect every living creature, but it was the only way for me to analyse and study the ant's behaviour in and on the water. I purposely did my research with red forest ants instead of flying ants. My reasoning was quite simple. When I built up my knowledge about river fishing in the mighty Scandinavian rivers I taught myself to observe the surface and surroundings continuously. During windy days, I often discovered red forest ants floating past me on the surface while I was wading. They even merged together in clusters, probably to prevent themselves from drowning. Aphids do the same thing. How they get into the water is still a mystery to me, but I strongly believe that the wind has a great influence. Fish took them like crazy and their aggressive feeding behaviour forced me to try ant imitations at an early stage of my flyfishing career.

Anyway, these special investigations at my tying desk finally led me to some striking discoveries and made it possible for me to design and tie the most lifelike ant imitations. It was not an easy process, and it took me several weeks of study and years of flytying development and fishing experiments, but the resulting dressings ended up very realistic and extremely effective.

Let's talk a bit about my observations. I quickly learned that a group of ants floating in the surface layer of a bowl of water show you more interesting things than you would ever expect. Good observation is unbelievably important for a creative flytyer and during my research I discovered that as long as ants float, they always hang very deep in the surface. I never saw any parts of their body clearly visible above the surface. So why do so many ant patterns float so high on the surface when the real insect is actually floating *in* the surface, or just below it? I think the answer is that most people want to see their flies clearly, and because most flyfishers and flytyers are mainly fixated on *flying* ants.

I also prefer my flies to be clearly visible on the surface, not only because my distance eyesight is a bit poor, but because sometimes fish just suck an insect from the surface with an almost invisible rise. This happens frequently when using flies like the Klinkhåmer Special or any other deep hanging emerger, and for me, a large wing serves

A nice grayling took one of my huge ant imitations

Yellowfish caught with a big size 6 ant imitation at Sterkfontein dam in South Africa

as a wonderful indicator for that sort of cryptic feeding and hard-to-detect rises.

However, for ant imitations, I make a big exception. When fishing with hardly any wind, there is no need to see them clearly on the surface, because fish often take ants so aggressively that it is almost impossible to miss the rise or take, and this is for me another example of the power of the ant. In stillwater, I fish ants while retrieving my line which can also explain their aggressive takes. The retrieving shows you where the fly is located as well. When you fish with artificial ants in lakes and big rivers where winds create a lot of waves, you certainly will need a good indicator on your flies to be able to see them in the turbulent surface water.

Huge legs

When ants land in the water it actually takes a long time before they finally drown, so I believe most of them will be eaten long before they die. In my tests, the minority of the ants curve their bodies, and when drowned, most just hang with their abdomen on top. It is very important to know that when ants are floating in the surface, their legs are extremely long compared to the body, and I really mean *much* longer than you see their legs tied in realistic and non-realistic dressings.

In their fight to stay alive and to survive, the movement of the legs is essential, and must be very attractive to fish. By the way, did you know that about one-third of the ants will finally sink after they drown, and that some of them will come up slowly to the surface again within a very short time? This is useful to know when using ant patterns in lakes or very slow running waters. It can help you while trying certain fishing techniques such as fishing a wet ant pattern on the drop, or imitating a surface-moving body with a high-floating pattern fished on a fine but sinking leader. In turbulent water this means that even drowned ants will look very lively, because the current will move the legs up and down and in all directions. This intensive research on ant legs resulted in me never tying any ant, or other imitations, with stiff legs made from materials that hardly move. Of course, it gives most of my ant patterns a much uglier appearance, but all I want to design are flies that catch fish instead of fishermen.

Why fish take the ant imitations so aggressively

In my way of flyfishing, I have a few singular theories, and, of course, not everyone agrees with me. That's what makes flyfishing so interesting and anglers so different. My strong belief in certain theories makes my confidence in most of my patterns extremely high. Whether some of my theories are correct or not, I really don't know, and I don't care either. As long as it helps me, and as a result, I succeed, I feel good about it.

One of these theories is that I believe that fish have a strong 'feeding memory'. I am totally convinced that fish recognise and desire certain foods simply because they like the taste of them. Let me give you an example: more than once, I discovered that during autumn fishing, fish simply refused every pattern, while going mad for aphids that get into the water on falling leaves or get blown in by wind. Aphids must produce a certain feeding memory that fish like very much. With ants it is similar. Did you ever try to taste aphids or certain types of ants? I did, and I can tell you that some of them taste extremely sweet. So maybe this is like 'filet mignon' for fish. For me, this is the reason why fish feed so aggressively and so selectively on ants and aphids. The difference between ants and aphids is that fish usually feed on aphids very sedately, almost invisibly, but take the ant mostly with a lot of aggression. This behaviour tells me that they don't want to miss their large juicy steak!

I also believe this is why Atlantic salmon take dry flies so eagerly after they have stayed a while in a river. The longer they are back in freshwater, the stronger their feeding memory returns!

One of the Ugly Bug prototypes, this one without extended body. Notice the extremely long legs!

Shape or silhouette

Another great idiosyncrasy of ants is that their special shape gives a marvellous and probably an easily remembered and recognisable silhouette that fish easily identify. In early spring, when the sunlight starts to get brighter, I fish a lot in Holland. My favourite places are some water-filled sand and gravel quarries not far from my home. Some of these quarries have huge sand banks, covered with just two feet of water. When the sun warms these shallows, my favourite time for flyfishing in Holland begins. Fish start to hunt for food more frequently, and the shallow banks become their most important feeding grounds. In March and April, there are not many people around, and the water is still crystal clear.

In the past, I have used mainly nymphs to tempt the feeding fish, but for the last 25 years, the ant and my Ugly Bug (a similar idea to the Chernobyl ant) have mainly done all the damage. It was in these local waters that I discovered the power of the ant well outside the season. The fish that taught me this lesson was the ide (*Leuciscus idus*). It is a wonderful coarse fish, highly appreciated by Dutch flyfishers because of its great surface activity, amazing size, fighting powers and great love for large dry flies or terrestrials. Thanks to my extensive flyfishing for ide, (Arctic) grayling and studying their amazing feeding behaviour, I was able to keep improving my ant patterns for such a long time. I also feel certain that yet more can be done in future to find the ultimate ant imitation. I especially refer here to the durability of my patterns by trying different materials.

Tying ants (the materials)

In the beginning, I only tied very simple floating ants. Since my research, however, I often fish with wet ants as well. A number of really large, wet ant imitations have been extremely successful in my personal choice of the flies I use in Canada and Scandinavia.

Wet ant made entirely from acetate floss

Sinking ants

One of the best flytying materials to make wet ant patterns with is acetate floss. Although acetate floss never became hugely popular, it's a wonderful flytying material, and highly underused by many flytyers. I once gave a tying demonstration in Germany on my technique for tying wet ants. Immediately, everybody wanted to buy and experiment with acetate floss, and the shop where I did my demonstration was sold out in just two hours. I prefer acetate floss because you can make the perfect ant silhouette in just seconds while you can easily merge soft leg materials into the body with very little effort. All you have to take care of is that your leg materials will not be dissolved or weakened by acetone.

For the ant it is only a matter of choosing the right colours of floss, and having a small bottle of acetone nearby. When I build up the nicely shaped body sections, I simultaneously include some legs (mainly made from pheasant tail fibres or similar, or rubber legs) in between the floss layers. As soon as everything is ready, I just add acetone to obtain the best shape, perfect colours and to harden the body. The final product is a simple, effective and extremely durable fly. The only drawback when using rubber legs is that rubber is highly affected by the influence of ultraviolet light and that's why I don't make too many at once and always store them in pitch dark fly boxes. I usually tie all my flies with rubber legs during my preparations for a specific fishing trip

Barbel took the Ugly Bug

Sometimes barbel can be very keen on ants too; this one took a flying ant imitation

Wet ants made from acetate floss

Yellow fish love big ants, especially the Ugly Bug

Lenok trout in Mongolia love big ant patterns and big means at least size 6

or one season only! Sinking ants, I fish a lot on the drop, which means the fish often take them while the fly is still sinking.

Floating ants

For my series of floating ants and Ugly Bugs, I mainly use four materials. These include foam, rubber legs, poly yarn and a hackle. Over the years I have made many variations and shapes, but I have always caught my larger fish on my simplest and most ugly dressings with the longest legs. I love my floating ants tied with rubber legs and why should I make my fishing flies complicated when simple flies will work just as well. My floating ant is one of the few patterns for which I choose a fragile material over the durability of the fly and that's because they are so quick and easy to tie. There is so much foam material available on the market today and I can only tell you what I have found most effective and convenient.

Soft foam (with a low percentage of ethyl vinyl acetate) is probably the most common flytying foam. Evazote is a low density, lightweight product that is closed cell, which makes it very flexible and softer and less dense than the 2mm variety often sold. It is resistant to several chemicals and UV light which makes it a perfect material for flytying. Polycelon, and Fly Foam, are other names for Evazote. I find it the easiest to tie with and it has better floatability because it has more air bubbles inside. When a more compressible material is desired, Evazote is probably the best. My favourite foam material is therefore not the durable craft foam that is used by many flytyers, but foam that is much softer, like Evazote. Personally I like Super Fly Poly Foam the most, which is Evazote. It is available in different colours and ready-to-tie strips so you don't have to cut any strips at all. A lot of flytyers prefer craft foam because it is so cheap and available in dozens of colours. In spite of the fact that it ties very well too, I still don't like it as much as Evazote.

For all my ants, I give the highest priority to the silhouette of the imitation. This is why I also tie a lot of my ants on curved hooks. I keep experimenting with different colours and body shapes and sometimes a slightly dissimilar body can make your day. Most of my ants are very easy and quick to make, so I don't mind if one breaks apart after several fish.

I will demonstrate for you a few different tying techniques that I use for my most simple parachute version of ants. Some of my dressings can be made in one or two minutes. Please note that I mostly tie the same ant patterns on straight as well curved hooks. I sometimes also use some extra long hooks like the Daiichi 1280 for making longer bodies or using the Daiichi 1260 when I want to fish for larger species such as brook trout, steelhead, salmon and wild rainbows.

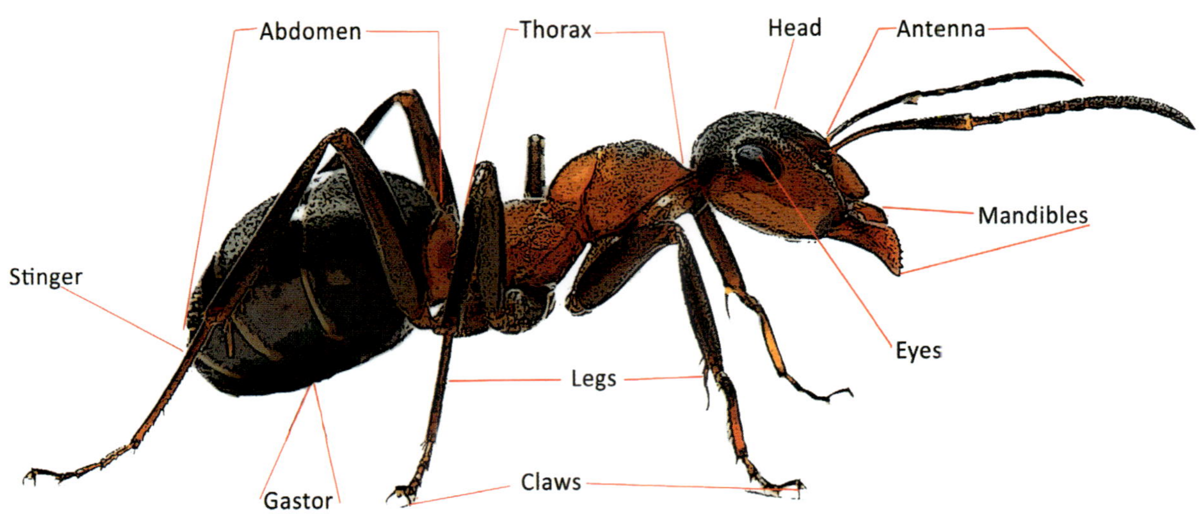

A simple ant anatomy, which I created graphically: all legs and wings are attached to the thorax

Round rubber legs are ideal for making ant legs in flytying

My preference: ready-to-use Super Fly Poly Foam

Semperfli wing posts or foam cylinders for parachutes that I use as a sight indicator on ants

Foam sheets just in case you cannot find the ready-to-use strips

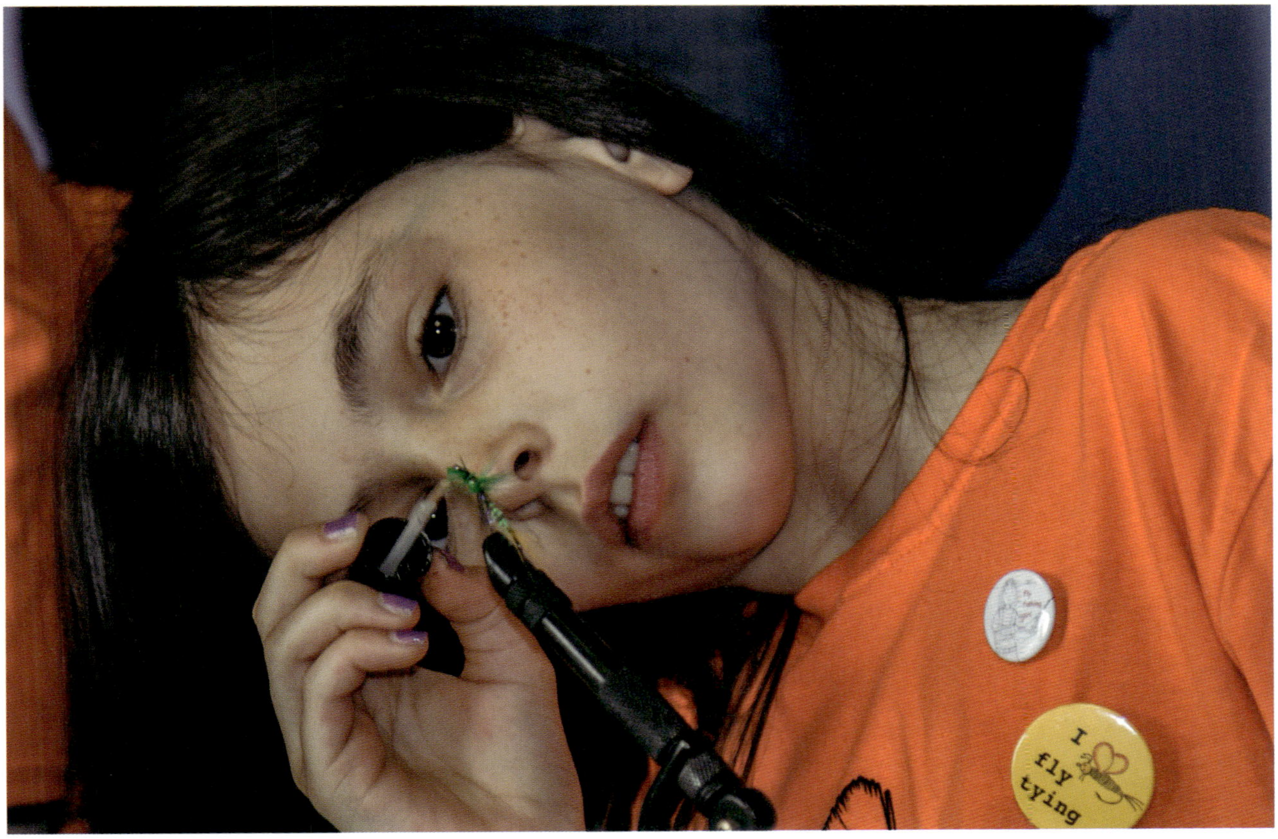

Kids love tying ants

Tying the one minute Parachute Ant (no 1)

Dressing

Hook: Daiichi 1180, 1182, 1280, 1260 and 1160-1167 sizes 16-8

Thread: Semperfli 8/0 Classic waxed thread in black colour

Abdomen, back and head: A strip of Super Fly Poly Foam

Legs: Dun-coloured hackle tied as parachute

Notes: This pattern is for slow running and stillwater.

You can also create your own strips from black foam sheet. In that case take 2mm thick Evazote foam and cut off strips of approximately 5mm wide and 6cm long.

Tying steps

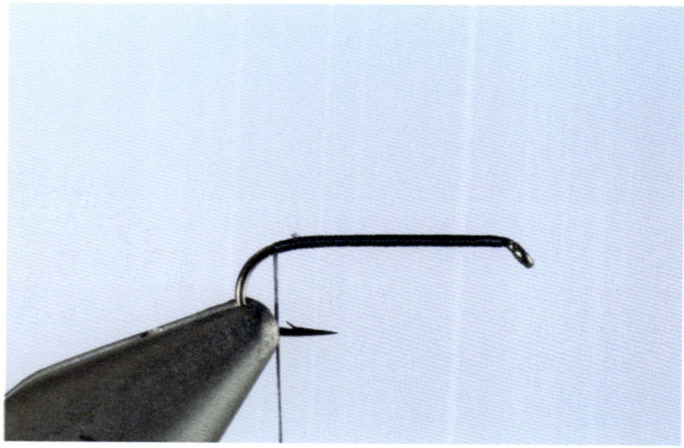

1 Place a hook in the vice and cover the shank with tying thread. This will help to secure the foam on the hook shank.

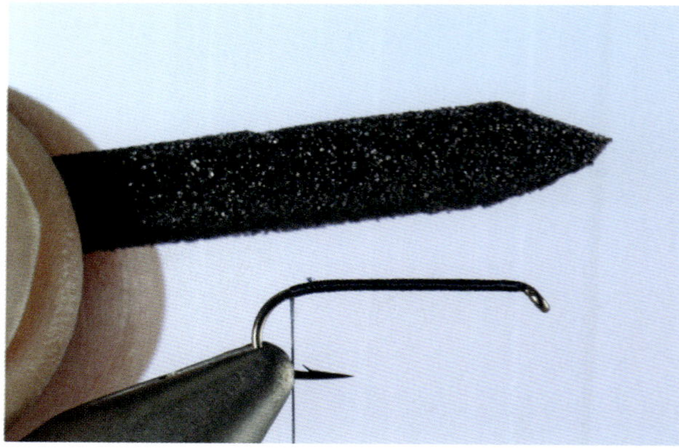

2 Pick a strip of Super Fly Poly Foam from your sheet and taper one side.

3 Bring your tying thread to the front and tie in the strip at the tapered side with very tight wraps from the front to the back and as far as possible.

4 Secure the foam strip well by making a nice tense underbody so as to help prevent the body from twisting around the hook shank later.

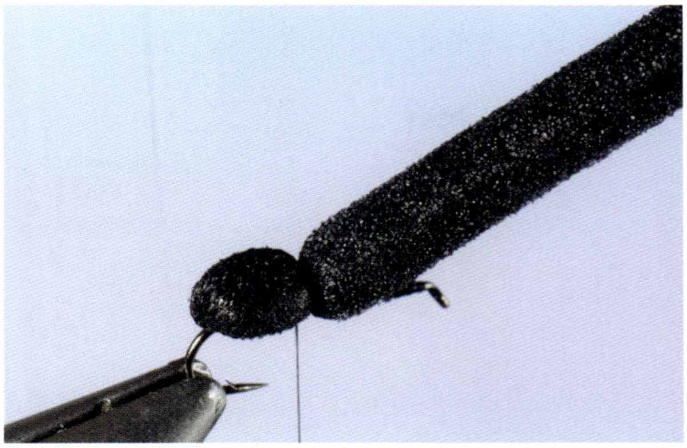

5 Now bring your thread back in the direction of the hook eye about 8mm from the bend to form the ant's abdomen. Then pull the foam strip over the hook shank and secure it with your tying thread.

6 Continue by making a nice ant's thorax by tightening up the foam with your tying thread in the direction of the eye of the hook again. Officially the thorax has to be the same length as the abdomen but I vary it.

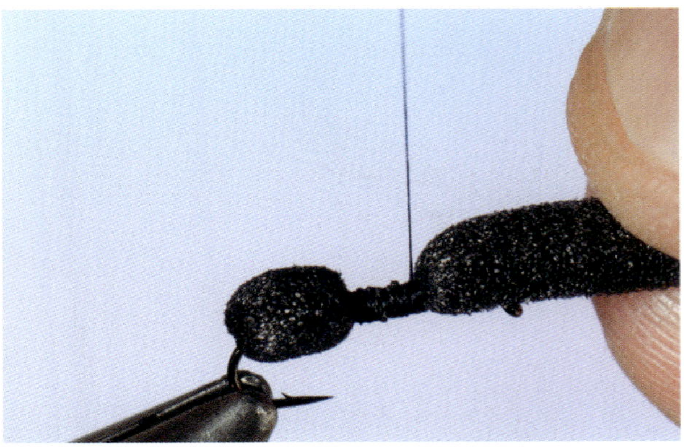

7 Don't cut the foam strip and be sure you leave at least 3-4mm space to make the ant's head later.

8 Pick a suitable hackle from your cape and remove the webbing from the quill.

9 Tie in the hackle by the quill in front of the foam strip, just behind the hook eye, with the tip pointing towards the bend. Cut off any extending quill and leave the thread before the foam strip.

10 Wrap a parachute hackle under good pressure around the foam strip.

11 Secure the hackle in front of the foam just behind the hook eye.

12 Pull the just-formed parachute fibres backwards between thumb and forefinger when you secure the hackle tip just behind the eye of the hook.

13 Now pull the foam strip towards to the hook eye (over the parachute hackle) to create the first part of the ant's head.

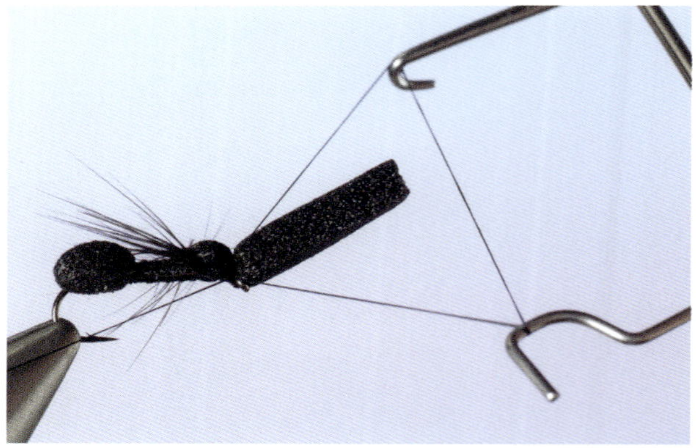

14 Secure the foam well with your tying thread, whip finish and cut off your thread.

15 Cut off the strip about 1-2mm long to form a bigger and oversized extended head. This extended head version can increase the catches enormously in some waters.

16 Add some lacquer to the windings with a needle or varnish applicator and the fly is ready to use *(experiment with the head sizes)*.

This bunch of large ants I took with me to Mongolia

Brown trout can't resist a well-presented ant imitation

Ant with and without a simple sight indicator

Tying the Parachute Ant (no 2, river version)

The dressing is exactly the same as described for Parachute Ant number 1. However, this version has a double layer foam as head and a little back part of foam which means this Parachute Ant no 2 will have a slightly different silhouette. This dressing is designed for waters with more movement and that's why I call it the river version. **Please note that tying steps 1-13 are exactly the same as I used for Parachute Ant no 1, so I will skip the first 12 steps in this tying session.** The little extended piece of foam that I tied purposely behind the head often holds a bubble of air underneath which creates extra attraction.

Tying steps for version 2

13 Starting position for Parachute Ant no 2 is the same as tying step 13 from the previous pattern.

14 Pull the hackle fibres forward with your fingers, then pull the strip backwards again over the first head, and secure the foam as shown in the picture above.

15 As you will see, you actually create a version of an ant with a much bigger head.

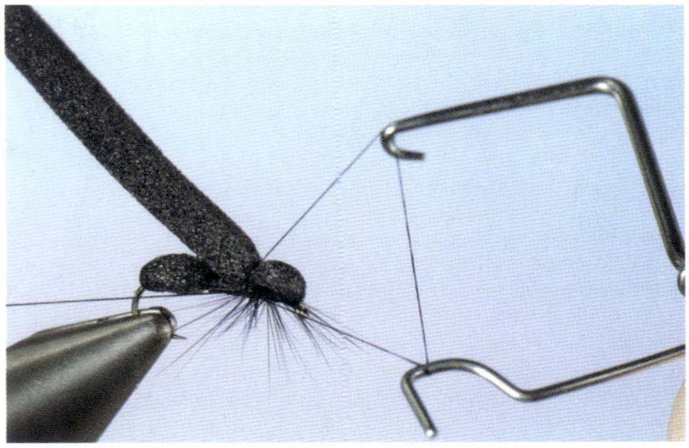

16 Tie off with your whip finisher.

17 Cut the strip just after the wrapping.

18 Completed fly.

Parachute Ant no 2 with sight indicator

Adding an extra sight indicator

With this simple technique I will show how to add a little indicator to your ants. When you fish ants at distance, or when wind conditions create a lot of waves, you will definitely need an indicator on your flies. Therefore, I tie at least 40% of my ants with an extra piece of white foam or polypropylene yarn to make them more visible. You can buy prepared soft parachute post or foam cylinders in many colours, or make them yourself by pressing a small copper tubing through a piece of soft foam.

1 For flies tied without an indicator, just set up your thread again, and tie in a round foam cylinder or a piece of polypropylene yarn on top of the head.

2 Now pull the cylinder deep into the foam with your thread. Move the foam into an upright position and tie off with your whip finisher.

3 Cut the indicator to the required size.

4 Completed fly with a sight indicator.

Flytying demonstration in Kuala Lumpur, Malaysia in 2012

The Ugly Bug – my version of the Chernobyl Ant

After a few years of intensive tests with parachute ant variations, I slowly started to add more and more rubber and switched to more Chernobyl-like ant variations around the late 1980s and early 1990s. This largely came about because of my passion to tempt the Ide in northern Germany and Holland during spring and early summer. During this period, I was able to catch them on very large ant and beetle imitations. Some of these patterns I allow to dance on the surface in order to entice the fish as much as possible.

Since 1986 I have barely missed fishing this prime time of the year and most of my terrestrial and beetle imitations and fishing techniques are based on the amazing surface-hunting strategies of this magnificent fish. My intense fishing in Scandinavia also encouraged me to use a lot of ant patterns. The more I experimented, the crazier my patterns started to look. When I got the unique opportunity to study and fish for the more aggressive Arctic grayling in the Yukon Territory and Alaska, I rapidly had enormous success using a series of ant patterns I called my 'Ugly Bugs'.

Although I have many variations of Ugly Bugs, and to avoid any misunderstandings, I will only share with you my most successful Ugly Bug pattern.

Set of Ugly Bugs as used in Mongolia, Canada and South Africa

Tying the Ugly Bug (the best of the best)

Dressing

Hook:	Daiichi, 1280, 1260 and Klinkhåmer hook 1167 sizes 16-8.
Thread:	Semperfli 8/0 Classic Waxed thread black
Body extension:	Super Fly Poly Foam 2mm (yellow, red, orange, green, black)
Abdomen, back and head:	Super Fly Poly Foam 2mm and peacock herl
Legs and antenna:	Rubber legs black
Indicator:	White soft foam cylinder

Tying steps

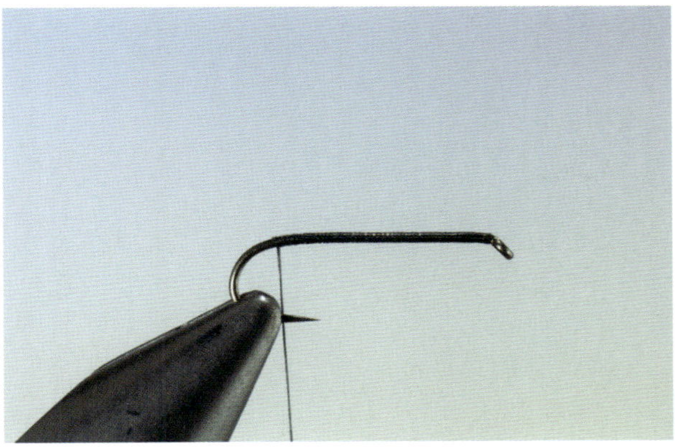

1 Place a hook in the vice and cover the shank with your tying thread. This will help to secure the foam on the hook shank.

2 Pick a strip of Super Fly Poly Foam from your sheet. Here I used a razor blade and ruler to cut my own strips from a piece 2mm thick Evazote foam sheet.

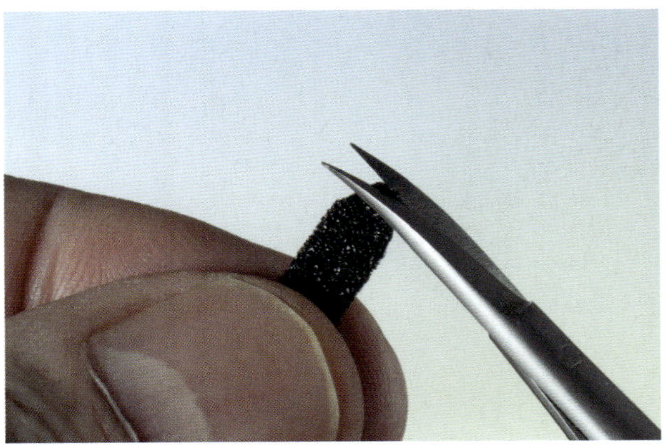

3 Round the edges with scissors as shown.

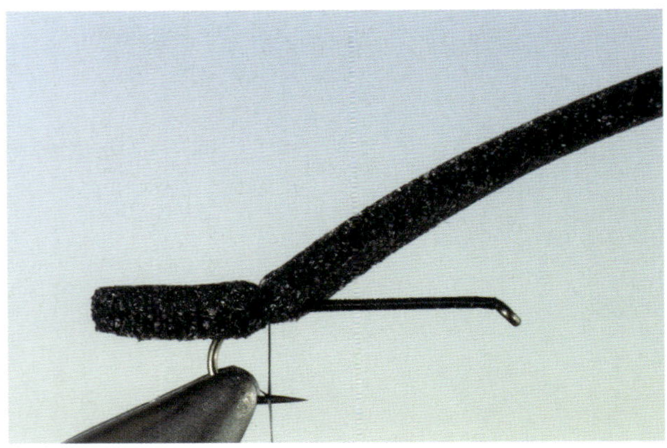

4 Tie in the extended body as shown.

5 Tie in 2 or 3 strands of peacock herl.

6 Wrap a small underbody with the peacock herl and trim off the excess.

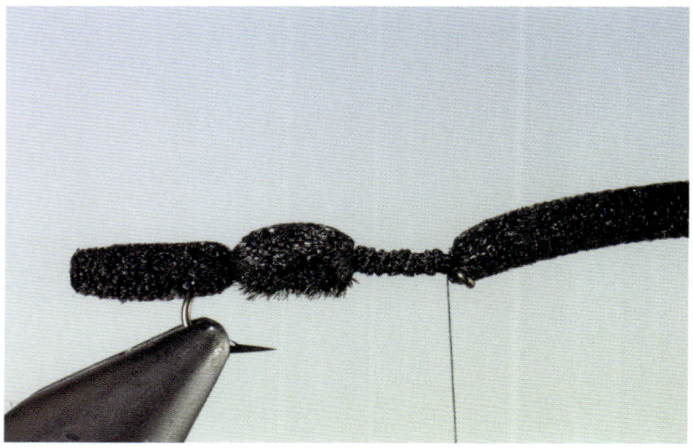

7 Pull the foam over the body under tension and then secure it well all the way to the eye of the hook.

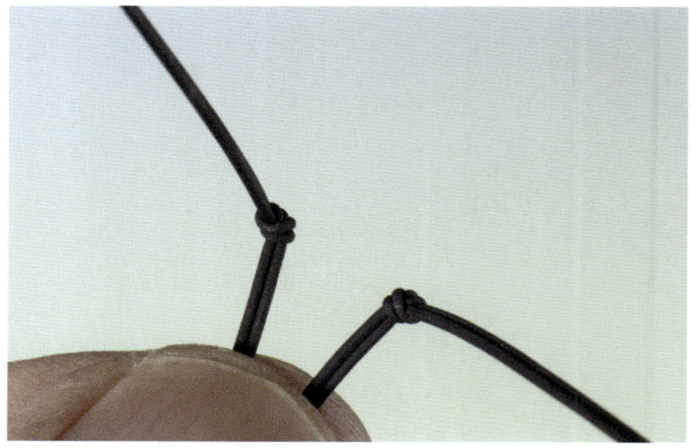

8 Take 4 round rubber legs and make two sets of identical legs as shown above, each leg contains 2 rubber legs.

9 Tie in the right leg and antenna over the whole length of the thorax. Be sure the bend turns to the inside and position the leg and antenna as shown in the picture.

10 Do exactly the same for the other leg and antenna. Don't cut anything!

11 Tie in at least two peacock herls.

12 Wrap a peacock herl thorax underbody again.

13 Bring the foam between the antennae to the back and secure it just behind the thorax in front of the legs. Cut the extended foam.

14 Tie in your sight indicator as shown earlier and tie off the fly with your whip finisher.

15 Cut matching pairs of indicator, legs and antennae to the appropriate lengths. You can even split the legs or antennae from which the rubber parts were still merged together.

16 Completed fly. (You can cut away two antennae if you like, and shorten them!)

Ugly Bug without extended body part and with added extra poly yarn and a large colourful sight indicator

Arctic grayling grabs an Ugly Bug

Ugly Bug with sight indicator

TINSEL
10 Yds.
MEDIUM

CHAPTER 10

Green Glomma Duns

During the summer of 1971, when I had only limited flyfishing experience and used extremely poor equipment, my casting wasn't very effective when trying to get out line. I was kept far busier untangling line, leader and tippets than trying to land my flies properly on the surface. Despite all the difficulties it was thrilling and I had become unbelievably enthusiastic after catching a few trout on an artificial fly in the River Mandal in southern Norway.

One year later, at the age of sixteen, while my friends were trying to impress the girls with their motorbikes in Luxembourg, I decided to travel alone up north to fish more beautiful rivers in Norway, Sweden and Finland. I had visited Scandinavia with my parents since the age of nine, but when you go all by yourself for the first time it is totally different and much more exciting. My father had bought me an Interrail pass so I could travel free all over Europe for one full month, and my plan was that a second ticket would bring me back home again at the end of the summer.

I left Holland around 15 June 1972 and after several days of continuous travelling, I arrived in the beautiful town of Dombås in Norway, where I found a lovely spot to put up my tent on the banks of the River Lagen. It was only thirty minutes' walk from the town centre but it made me feel as if I was deep in true wilderness country. Here I met two pretty young outdoor ladies who taught me a few of the skills needed for flyfishing and with whom I had a long and unforgettable midsummer night party on 21 June.

I fished the area for eight full days and when the time came to leave this wondrous place and move further north, the older of the two girls gave me her almost brand-new rod and the younger one gave me a reel, a line and a box of flies. I was so thrilled and excited that I forgot my father's old cane rod and left it standing against a tree at my wilderness campsite. Thinking about it now, that rod may still be there.

Living in unspoiled wilderness

As a high school pupil with plenty of spare time and some extra pocket money from my father as payment for helping him in his business, I managed to spend most of my summer holidays in the northern parts of Norway, Sweden and Finland. These were the beginnings of my eternal love affair with the great outdoors. In 1975, I stayed for almost four months in the largest 'unspoiled wilderness' in Europe: Northern Finland, in an area better known as the land of the Sami people. Here I learnt how to survive in the deepest forests and on the tundra.

My teacher was an old Laplander and we communicated by pointing and signing. It was incredible how much I learned about nature and wildlife, about true friendship, and about myself. I learnt more than all my schools had taught me in my short life. It was amazing to see how removed we have become from the hunter-gatherers of our not-too-distant forefathers. The holiday came and went all too soon; this was to be my last, as my student days were over, but I kept heading off north on every opportunity I got.

The beautiful Glomma river where I learned so much about flyfishing

An over 50cm black grayling caught by Hein van Aar in the Secret river (©Hein van Aar)

German times

In 1976 I had to join the army and started my education in the armed infantry to qualify as a sergeant by the end of the year. In 1977, I was sent to Germany where I did my obligatory military service. During my job, I had to replace my professional platoon sergeant for some time and executed my duties so well that my bosses asked me to stay on and motivated me to join the army permanently. I know I must have disappointed my father enormously in not taking over his business, but my mum forced me to choose the army, promising a much better and more secure future with good promotion prospects and a stable income. She was right and I never regretted my decision because when I retired from the army in 2012 I was commander of the armed infantry gunnery school for heavy weapons.

The 1980s and early 1990s represented a period in my life in which I had a few very busy and responsible army jobs. It was hard work, but I had plenty of spare time to build up my flyfishing and flytying skills. During these years I had many unforgettable outdoor trips in which I not only developed a few dozen fly patterns, but also tested countless flies for their floatability, durability and effectiveness. During this time, I married my girlfriend Ina and found perfect harmony in my marriage and with the wild. I also did a large amount of experimental flyfishing and discovered, and created, a number of fishing techniques, such as my special combination of a parachute and snake cast, long downstream drifts with dry flies, and made my own braided leaders and my knotless connection system between the backing, fly line and leader. These are just a few of the many ideas I had during those wonderful and productive years.

My third army job was again in the north of Germany; remember, the Cold War was still on, and my work was based on 24/7 duties for many weeks in a row. Then I was offered excellent compensation in terms of time instead of money, and that was the main reason why I chose that

My first Green Glomma Dun prototype with CDC wing and tied on Swedish dry fly hook (1990)

particular job at that specific location. A lot of colleagues left their wives back in Holland so they had to go back each holiday to keep the family ties close. Ina, however, had decided to live with me in Germany and we had a great time. I also soon learned that there is absolutely nothing wrong with having a busy job when there is time to recharge your batteries, especially if I planned my holidays cleverly. It brought my life into perfect balance; at least that was my personal experience of it.

Second home

From our new home in Germany it was only a couple of hours drive to Denmark, which created the unique possibility of travelling even more frequently to Scandinavia during the many years that I was stationed there. Traffic jams were rare in those days and these trips meant additional fishing too. Ina loved the outdoors as much as I did and we could therefore spend all our spare time together in the north. She didn't flyfish until 1991, but she enjoyed watching me, reading books or taking photographs from the riverbank. This extraordinary experience made the 1980s not only my golden years in flyfishing, but also my most productive years in flytying.

I fished many years in Denmark, Norway, Sweden and Finland, but my most successful and best fishing in Scandinavia was without doubt 1988, when I had four major breakthroughs in my flyfishing career. (*See the chapter on my Poly Bugs.*)

When I recall my Scandinavian fishing trips my heart immediately starts to beat faster. The moment I arrive in Denmark, Norway, Sweden or Finland I feel as if I am coming home, and it really hurts when I have to leave again. I think these feelings sum up how much I like these wonderful fishing paradises and to meet up with my friends each year. Today I have very similar feelings for Atlantic Canada, the Yukon Territory, South Africa, Mongolia and Malaysia as well. I think it's very normal to have feelings like this when you have spent very long holidays fishing and travelling in the same places. In the 33 years I have been visiting Scandinavia in a row, Denmark, Norway, Sweden and Finland have become my second home. When I think of Norway I see the fjords, the mountains and the mighty rivers as clearly as if I were actually standing there. When I think about Finland I see the amazing midsummer nights and the midnight sun which have always fascinated me, and they provide the welcome bonus of many extra fishing hours. My circle of friends so high up in the north is incredible. When I think of all the fishing, it's the 'Lady of the Stream', the grayling, that has made the biggest impression on me and my ties with her are very strong and personal.

Of course, there are hundreds of rivers in Scandinavia that produce excellent fishing, so let me tell you why Norwegian grayling in particular drive me so crazy. It all started in 1981. I was fishing the River Sömåa in eastern Norway when I caught my biggest-ever grayling on the Rackelhanen. It was a beautiful fish. I have never broken this record since, and although I have matched that fish's unbelievable 61cm length twice more, in 1984 and 1989, it is the memory of my first `black grayling' that is strongest in my mind. I had never seen such a large and dark-coloured grayling before. She was well hooked and started to bleed when I removed the hook. However, when I released her, she turned her head as if to look at me before she slowly disappeared into the depths again. Whether she actually did look at me I still don't know, but that look made such an impression on me that it triggered my passion for grayling for ever since. I knew I must have hurt her, and from that moment I never used a barbed hook again.

How things really got into the swing

In 1977, when my first army job started, I could no longer spend months in a row in the northern part of Norway, Sweden and Finland, and my

Isterfossen, often claimed to be the source of the River Trysilelva/Klarälven but the real start is the outflow from Lake Rogen

Beautiful dark coloured Scandinavian grayling caught on opening day in Norway 2013

One of many special grayling rivers that I discovered by walking long distances in the 1980s

holidays had to be shortened. This meant I had to concentrate my flyfishing more on Denmark and the south and the middle of Norway. For three full seasons, I went into research mode and finally made a very important decision: to start focussing mainly on two huge watersheds, both arising in the beautiful county of Innlandet in Norway.

In the most eastern part there was the Trysilelva (Klarälven) river system in Sweden, including all the headwaters, tributaries and lakes that feed it. Further west you find the entire Glomma river drainage basin, the longest river in Norway, which I started to explore from the town of Koppang only two years later. Both river systems are not far from each other and I sometimes fished both areas in the same season. In other years, I fished them completely separately. I followed both river systems upstream to their very sources and I fished their watersheds intensively until 1996.

I always combined my main summer trips with two weeks of grayling and brown trout fishing and then two weeks fishing exclusively for Atlantic salmon and seatrout on the west coast of Norway. During shorter trips, I often went after salmon or grayling only. It was in Innlandet that I discovered a few important 'rules' that became the key to my success and knowledge for grayling and trout fishing and their feeding behaviour worldwide.

One of these rules was to cover long distances on foot along riverbanks and through the forests. Do the mileage, and search for places without human footprints and you can be sure you will find your hotspots.

The majority of fishers visiting Scandinavia aren't into long hikes and they prefer to grab their car and just fish the easily accessible water, so the more miles you hike, the better your chances of catching real whoppers.

My strong natural inclination for exploration with the aid of good topographic maps and my army background made most outdoor activities a lot easier for me. There was no GPS as we know it today and I had to work with a compass and maps all the time. I have no idea how many rivers I swam across during my journeys, but it must be a few dozen. Some of my hotspots were just small connecting rivers between lakes about two or three hours walk from the nearest road.

One of my favourite spots includes 22 kilometres of top grayling water and is only reachable after a four-hour hike. The place is still good today and a few close friends of mine still fish it every year. In my diaries, I have described seventeen spots like that: places where I usually spent a few days on my own bringing a little tent to store my backpack, food and equipment, or take a nap in when I got tired. Of course, you have to know exactly what you are doing in such remote areas because 90% of these hotspots are most successful during the first two weeks in June. Later in the season the fishing declines rapidly because the bigger fish move back into the lakes again. The end of May can be very good, but you have to check the opening days for any particular area. Going too early can also be risky due to the unpredictable weather, like icy mountain winds or even snowstorms, and neither are great experiences, believe me!

After numerous fishing trips over a period of several years, I began to discover some of the best grayling and trout spots in the headwaters of the Trysilelva/Klarälven system. In the old days rivers like the Sölna, Sömåa, Hola, Röa, Mugga, Femunsdelva and Glöta were like the Garden of Eden for every grayling angler, and lakes like Femund, Sölensjöen, Aresjöen, Isteren and Feragen were just a few of my many newly discovered paradises in the 1980s. Most of these waters are between 600 and 1,000 metres above sea level, the weather conditions are tough, and the fishing season is short, although very productive. There is also an enormous population of wild and

Look at the shine and glitter, produced by the broken surface water (©Hein van Aar)

hungry mosquitoes, black flies and no-see-ums, and you just have to learn to deal with these before you can find your enjoyment. This area offers grayling what they like most: lakes with excellent conditions for survival during the severe winters, and have an abundant food supply; the insect life is prolific, with superb caddis hatches and many species of mayflies. Most of the streams in the area are very similar, though they vary greatly in length, width and depth.

My 'Shining River' (The Glomma)

I explored the Glomma River in exactly the same way as I did the Trysilelva/Klarälven system, except a lot faster. About ten kilometres upstream from Tynset, you will hit the area that I call the 'Shining River'. From here all the way up to the outflow from Lake Rien, the Glomma can be considered the crème de la crème of grayling angling.

Trout fishing, however, can also be spectacular during the midsummer nights. This is not just my experience, because many books and magazine stories will tell you exactly the same. It was in this area in particular where my good friend Hein van Aar was able after many years to achieve what I couldn't. He continued where I had stopped and failed. I taught Hein how to tackle the river in my early days, and he has worked himself up to become one of the most renowned flyfishing guides and riverkeepers in the entire Glomma river system today. He is a true master who guides his guests with passion, exactly in the way I guided myself for so many years, and how I wished to be guided in foreign countries too. I only guided close friends and there was never money involved.

Hein is an awesome photographer and I am very happy that he gave me some pictures to share with you in this chapter. Together with the Tolga River Board, the Tynset River Board and some local farmers, Hein started Kvennan Fly Fishing in 2006, as a long-term project to preserve quality river flyfishing for the future.

They introduced the first fly-only zone on the Glomma River, and to protect the large fish, they even limited the number of rods, introduced a bag limit, and raised the minimum size for both trout and grayling. They work closely with the Evenstad University College Faculty of Forestry and Wilderness Management, monitoring fish health, fish stocks, water quality and several other factors of vital importance. In the fly-only zone there are many of my early hotspots, such

This picture shows clearly why I renamed the Glomma 'The Shining River' (©Hein van Aar)

Hein van Aar posing at the Lakes of Hodalen. (©Hein van Aar)

as The Bermuda Triangle, Elk Isle, Selund Isles, Hagen Isle, the Upper and Lower Labyrinth, Otter Island, and the nasty Neset Hole where every year at least a few dozen flyfishers have a good swim. I introduced some of these names in the early 1980s and they are still in use today.

Every year flyfishers from all over the world visit Kvennan Fly Fishing to wet flies in the fly-only zone, and I am unbelievably proud of Hein who did all this hard work to create a catch and release area a lot more effectively than I did. If you are a grayling addict you must have a look at Hein's stunning website at: www.kvennan.com and don't worry about the language. It is all written in English. On the site, you will find the latest fishing reports, information about climate, water levels, a lot of fishing information and stunning pictures. You can contact Kvennan Fly Fishing by sending an email to: kff@kvennan.com.

The river of inspiration

The best way to be successful is to merge with the river, find your harmony and look for the shine and twinkle effects produced by the always broken surface water. This whole area is not only beautiful, but also very quiet. There are just a few peaceful villages and towns, which you can easily fall in love with, and the people are extremely friendly and always willing to help you. It is in this area that I developed my big Light Tan Series of 44 grayling flies between 1982 and 1992, and several are described in this book. Each of them

Beautiful close-up picture of the male imago of the Leptophlebia cupida

First autumn snow, Glomma River (©Hein van Aar)

was designed when we camped out at Kvennan, or while we stayed in our wonderful cabin on the banks of my 'Shining River'. Big flies seemed to be the secret to catching the bigger fish, and that's why my Scandinavian patterns increased so much in size.

The River Glomma gave me enormous inspiration to tie flies and I had many new ideas and patterns to investigate and play with – time well invested. For example, the summer of 1990 was a very experimental one, especially for my grayling fishing, because I took home dozens of insects from my 1989 trips, which I evaluated and recorded during the winter of 1989-1990.

I had two particular goals in mind: the first was to create a good looking and effective mayfly pattern, which I could use all over the sub-Arctic regions, like a kind of a universal mayfly. My second goal was to design a fly that was entirely made from natural materials. This was because I had used lots of synthetics in my flies since I tied the Rackelhanen and so many Klinkhåmers. To realise my goals, I had to rely a lot on the notes that I made in my diary about mayflies and their hatches.

Entomology (keep it simple)

In this chapter I will give you a series of dun or spinner imitations that pleased me a lot after all the research and fishing that I did. In the flyfishing world, we refer to the sub-imago as the dun and the imago as the spinner. I also want to give you a little background about some of the adult stages in the life cycle of mayflies but in a way I see as the simplest. I will share with you how I went into the life cycle of insects and will try to give you a few essential differences between the males and females of adult mayfly species.

For a lot of people, entomology is complex, difficult and challenging, and you have to memorise a lot or walk around with identification charts. However, if you don't go into research and identification *too* deeply it can actually make things a lot easier. I tell you this because at a certain stage of my flyfishing, my biggest problem with insects was that I got so obsessed, that I wanted to know the exact species. That is not only extremely time-consuming, but almost impossible. You have to walk around with a microscope to determine an exact species, because often it is only through the close scrutiny of genitalia of the male imago that you get the answer. I thought identification would be a lot easier with excellent macro pictures in very high resolution, but it isn't, and that was another of my turning points.

The late Oliver Edwards, one of Europe's leading flytyers, enthralled me with his effective and realistic fishable fly imitations. We became close friends during all the fishing, classes, workshops and flytying demonstrations we did together in Europe and the USA. Thanks to my fishing trips with Oliver in the 1980s and early 1990s, I learned a lot about entomology and he was the first person who actually explained to me the facts and myths about insects in a simple and very understandable way. He made me realise that you don't have to dig too deeply into the exact identification of insects. Once you find a nymph and know the insect's family name, you don't need to go much further into its classification. Most insects within the same family have a comparable larva structure with very similar general characteristics. Adult mayflies are much harder to identify and if highly qualified scientific people sometimes struggle to define a particular adult mayfly, why should I punish my brain trying to do the same? A great help to me more recently were experts like Steve Burian from Connecticut State University and Stuart Crofts who is a tutor with the Riverfly Partnership in the UK. Stuart also does freshwater courses for the Freshwater Biological Association in the UK, to help anglers find their way around the world of freshwater insects.

Let me give you another example. Several years ago, I had found an adult mayfly in Labrador, which I identified as belonging to family *Leptophlebiidea*. I guessed it was a *Leptophlebia nebulose,* but after scientific research it turned out to be a *Leptophlebia cupida*. I was very wrong in the exact species but correct with the family

Hein van Aar with another beautiful grayling caught in the Gomma watershed (©Hein van Aar)

Picture of the sub imago female Danica with dull wings sitting on my Klink cane rod

and genus and I think two out three is not a bad score after all. For my flytying it was more than enough to recognise the genus as *Leptophlebia* because when I searched among the artificial flies in my fly boxes, I had an enormous success with an imitation named *Leptophlebia marginata* that I had developed while fishing the 'Secret River' in Norway. This was further concrete proof to me that you can indeed catch a lot of fish using an imitation of the genus or family the mayfly belongs to, while the fish is actually feeding on a completely different species. The brook trout in Labrador took my version of *L. marginata* as eagerly as if it was a real *L. cupida*.

Another important reason why adult mayflies are so hard to identify is that for every species there are four variations: male sub-imago, female sub-imago, male imago and female imago. Sometimes only one of these can be identified positively to full species level. For adult mayflies, I think identification is easiest for the spinners (imagos) because the mayfly is then sexually mature. When you identify adult mayflies, a lot of people look at size, legs, wings, hind wings, tails, colour and gender. For me it was an enormous benefit to know the differences between a male and female adult mayfly. I can say the same for understanding the life cycles of duns and spinners, and knowing these differences has been extremely helpful to me.

For those who are not so familiar with mayflies there are a few observable features that I use to determine the different sexes in the winged stages of imagines or spinners:

1. Males usually have much larger eyes than the females.
2. Males have elongated forelegs for grasping the females during aerial mating. (Female's forelegs are usual shorter.)
3. Males have forceps (claspers) at the end of their abdomens to hold the female during copulation. (Females don't.)
4. Males can have a much longer tail (Danica).
5. Females can be clearly larger in size (Danica).
6. Sometimes there can also be clear colour differences between male and female (Danica).

While not all these features will always be observable, usually two or three are, and can be used to determine the sex between a male or female of a winged mayfly. Please note that even in the dun stage it sometimes can be difficult to tell the difference between male and female mayflies.

Difference between duns and spinners

Mayflies are unique in the insect world in having more than one winged life stage, the dun and the spinner. Usually they moult twice within 24 hours! The final transformation is from a dun (sub-adult or sub-imago), to a spinner (adult or imago). Under most circumstances, the identification of duns from spinners is easy by looking at the wings really closely.

- Most mayfly duns (sub-imagos) have wings that are dull and opaque.
- Most mayfly spinners (imagos) have wings that are shiny and transparent.

Other differences between duns and spinners

- Duns are good at managing water. It's as if they have an extra waterproof layer, but they are awkward fliers.
- Spinners have longer legs and tails and are real acrobats in the sky.

Worth noting is that duns and spinners have mouthparts, but they are vestigial and therefore non-functional, so they don't bite. The dead or dying egg-laying female imago is the spent

Female Danica newly hatched from sub-imago to imago: note the clear transparent wings

Clear differences female (left) and male Danica

My check for male or female always goes: eyes, front legs, claspers, tail size, body size and colour differences

Beautiful picture of a female Danica with clear opal-coloured wings

Female Danica with dull opaque wings

Male Leptophlebia cupida determined by three observable features

spinner, same for males in species that swarm over the water.

So, in summary, the life cycle of mayflies is not that difficult to understand: egg to nymph, to emerger, to sub-imago or dun, to imago or spinner, and finally, to the spent spinner when dying or dead.

New way of thinking

Since the mid 1990s I changed my way of thinking and made new plans for my research into insects and my nymph imitations for the future. For the first time in my life I understood that as soon you were able to recognise a *Baetidae* larva among the many other nymphs, you already have more than enough information to tie a perfect artificial of the *Baetis Centroptilum, Cloeon* or *Procloen* nymph. When you get different *Baetidae* species living together they all have slightly different habitat preferences in the stream, but all these are in the realms of trout and grayling, so they will not be bothered which they eat as long as they can get at them.

The different *Baetidae* species also often have slightly different periods for emergence to avoid unnecessary conflict for food. To pick a *Baetis* species from the *Baetidae* family is not even that difficult because the middle tail is shorter for all *Baetis* larvae, but do you think the fish will notice? There are 14 species of *Baetis* larvae in the UK and I think it doesn't matter whether you have tied a realistic copy of the *fuscatus, scambus, niger* or *rodani*. I discovered that when you use the structure of a *Baetis rodani* nymph in your pattern you will also catch fish that are feeding on *Baetis fuscatus, Baetis scambus* or *Baetis niger*. When I was really deep into my amateur entomology research and was catching fish with a lot *Centriptilum luteolum* or *Cloeon dipterum* in their stomachs, I was actually using my *Baetis rodani* imitation.

Meaning that I found my solution in flytying: just tie a perfect copy of one of the species from the entire family and I strongly believe this works very well. For nymphs, I also find that the imitation of the exact colour is less important than it is for adults, but it's just a feeling I have. In any case, if you want the perfect colour match you must always check the colours in a dish of water when your nymph pattern is completed.

Perhaps colour matters more when flies float on the surface because fish see the adults from a different angle and against the light, but I don't know that for sure of course. I only know that colour is often a main key to success while dry flyfishing.

In this chapter I will give you a description of an adult mayfly series that has served me very

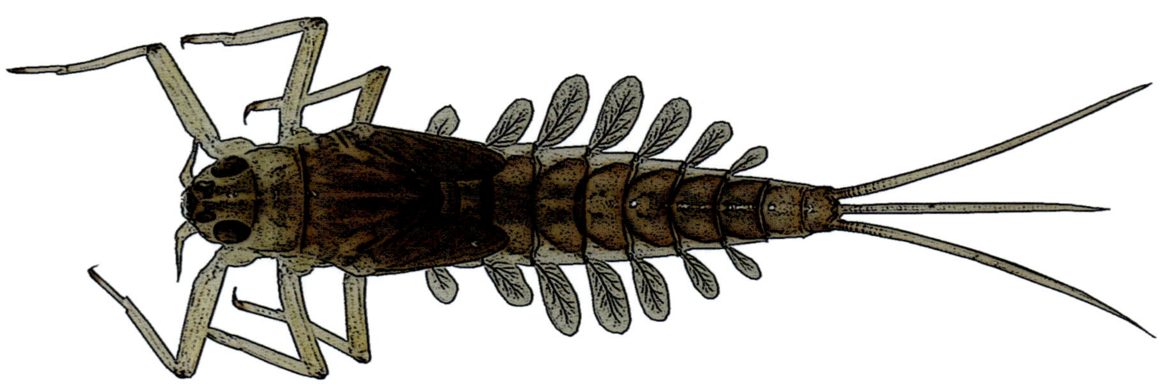

All the nymphs of Baetis species that I have found have their middle tail shorter (created graphically)

Even when you don't know the species, you can see clearly that this mayfly is a male because of its claspers

well since the early 1990s. I use them mostly for dun and spinner hatches in the sub-Arctic regions. I have fished these patterns over many years and at many different locations and have always caught a good number of fish on them, including during quite heavy hatches. I call the patterns, or series, the Green Glomma Duns. I prefer the term duns because it sounds a lot better than spinners. I went for 'Glomma' duns because I developed them in our cabin on the Glomma River, and I selected green but also use one of the most beautiful olive-green colours imaginable, which I got by dipping natural grey feathers in picric acid.

I designed these patterns especially for those occasions when trout and grayling are feeding selectively, because only under these circumstances did my Klinkhåmer patterns sometimes fail. I also wanted a pattern that was suitable for rivers and lakes, because in many lakes of Innlandet there are big hatches of mayflies. I put most of my effort into the development of a pattern that floated well and, especially, a fly that landed perfectly because I was so spoiled with my Klinkhåmer. I didn't find the solution immediately, but after two months of experimentation I came upon a marvellous idea by accident when I was selecting slides from my Light Tan Series for a lecture. A few of my slides were sitting upside down in the slide holder and when I looked at them more carefully I saw an excellent opportunity to try a different technique for tying the hackle fibres upwards at the bottom of my mayfly imitations. I prefer the idea of tying up the hackle fibres instead of cutting or burning them because it generates a kind of stabiliser in the fly's floatability. With this technique, you create a fly that keeps all the hackle fibres and therefore the pattern stays more voluminous, while at the same time it also has an excellent silhouette, and will always land perfectly on the surface. A fuller fly also drifts better in more turbulent water.

For years I used a technique that I learned from an old Swedish guy who showed me how to tie up the bottom hackle fibres in collar-hackled flies using the tying thread just before tying off the fly. It was not an easy technique and trout teeth often damaged the thread and the hackle fibres came loose. For the flies, I saw in the upside down slides that I had used a piece of herl and poly yarn to split their wings, so when the slide was upside down I got the idea to do the same thing for the bottom hackle fibres too. It was as simple as that! By using this technique, I not only had the same results, but a much more durable fly as well. A second benefit is that the fly is far easier to tie. I use this technique a lot for my large mayflies because they float so very well on the surface.

While continuing to experiment with this idea I included a valuable tip from Philip White whom I met for the first time in 1987 when he was still the riverkeeper for the River Wye in Derbyshire. Since then we met several times in the late 1980s and early 1990s and became really good friends. We also had many discussions about fishing techniques and our flies. I loved Philip's thoughts behind his flies, and they work so well, not just in Derbyshire and the UK, but in many places around the world. His special series of struggling mayfly imitations I still use regularly. The technique I learned from Philip that I want to show you in this chapter is tying the wing over the collar hackle which I had never seen anybody doing before. It's a very simple technique and extremely effective so, thanks to Philip, the Green Glomma Duns got even better.

The Green Glomma Duns worked well for me during hatches of *Ephemerella*, *Baetis* and even the *Ephemera* and *Lebtophlebia* species. The

The inspiration for the Green Glomma Dun (© Hein van Aar)

Three male sub-imago of the Leptophlebia cupida (Note the dull wings.)

The Glomma River at its best in early summer (©Hein van Aar)

Two early prototypes of my Green Glomma Duns tied with deer hair wings and blue dun hackle

My favourite spot for fishing the famous lakes of Hodalen

A CDC version of the Green Glomma Dun tied by Leon Links

most important insect that I tried to imitate with these patterns is known as *Ephemerella aurivilii*, a mayfly that mainly occurs in Scandinavia, Nova Scotia and Labrador. When you change the body colour, this pattern can be adapted to imitate many other mayfly species.

Forking the tails

To split or fork the tail I use a very nice technique that was showed to Al Beatty at the FFF 1989 Conclave by an observer and was popularised by Wayne Luallen, a good friend of mine. The trick is to use a scrap piece of thread to achieve the simplest way of forking or splitting tails. Since Wayne showed me the technique, I have used it in a lot of workshops and classes with great success. The same method works with three tails as well, as long you keep one strand of thread on either side of the middle tail fibre. (*See tying steps Green Glomma Dun no 3.)*

Tying the wing

It was Philip White who gave me the idea of tying in the wing over the hackle. I was so impressed by this tying technique that I made hundreds of them in three different variations and fished them with unbelievable success all over the world. Thanks to Philip I made a very nice series of mayflies based on his technique. The only difference in my variations relates to the wing materials, because with this particular fly I wanted to use 100% natural materials and that's why I use deer hair, duck body feather fibres and CDC as wing materials.

Today I also use non-dyed, natural greyish-black wing fibres for the body and a natural tan hackle for those situations when the fly's colour better matches the hatch. I call them the Dark Glomma Duns. The fly design and tying steps are exactly the same.

Original Green Glomma Dun series from 1991 (© Rudy van Duijnhoven and Hans van Klinken)

Dressings Green Glomma Dun

Green Glomma Dun No 1 (Used and extensively tested in the summer of 1990.)

Hook: Daiichi 1280 size 10-14

Thread: Semperfli 8/0 Classic waxed thread, black or olive

Tail: A bunch of grizzly fibres dyed olive (about 20mm long)

Body: About five fibres from a large wing or tail feather dyed olive

Ribbing: Semperfli pure silk vintage orange or yellow wrapped in opposite direction.

Wings: A bunch of greyish-brown duck body feathers split in 2 sections (about 20mm long)

Hackle: Grizzle hackle dyed olive. Tied up with fibres from a wing or tail feather

I took this photograph (*left top*) together with Rudy van Duijnhoven somewhere in the early 1990s. This was the first prototype of my Green Glomma Dun that I developed on the banks of the Glomma River in the summer of 1990. I still use the same pattern all over Norway, Sweden and Finland.

Green Glomma Dun No 2 (Used and extensively tested in the summer of 1990.)

Hook: Daiichi 1280 size 10-14

Thread: Semperfli 8/0 Classic waxed thread, black or olive

Tail: Two very fine dark brown tips from a patch of peccary (wild boar) hair or dark olive/brown coloured Microfibetts

Body: About five fibres from wing or tail feather dyed olive

Ribbing: Semperfli pure silk vintage orange or yellow wrapped in opposite direction

Wings: Two greyish-brown duck body feathers burned with wing burners (20mm long)

Hackle: Grizzle hackle dyed olive. Tied up with fibres from a wing or tail feather

Also photographed together with Rudy (*left middle*). It was my second version in the series developed in 1990. It is a beautiful fly to make and one of my favourites in my workshops when we do mayfly patterns with realistic wings.

Green Glomma Dun final version (Used and extensively tested in the summer of 1991.)

Hook: Daiichi 1280 size 10-14

Thread: Semperfli 8/0 Classic waxed thread, black or olive

Tail: Two very fine dark brown tips of peccary hair or dark olive/brown coloured Microfibetts

Body: About five to seven fibres from a large wing or tail feather (dyed olive)

Ribbing: Semperfli pure silk vintage orange or yellow wrapped in opposite direction

Wings: A bunch of greyish-brown duck body fibres

Hackle: Grizzle hackle dyed olive. Tied up with fibres from a wing or tail feather

Also photographed together with Rudy (*left bottom*). It was my final version in the series as tied in 1991 and my most successful one when it comes to numbers of catches. *That's why I present only this version in my tying steps overleaf.*

Tying the Green Glomma Dun (final version)

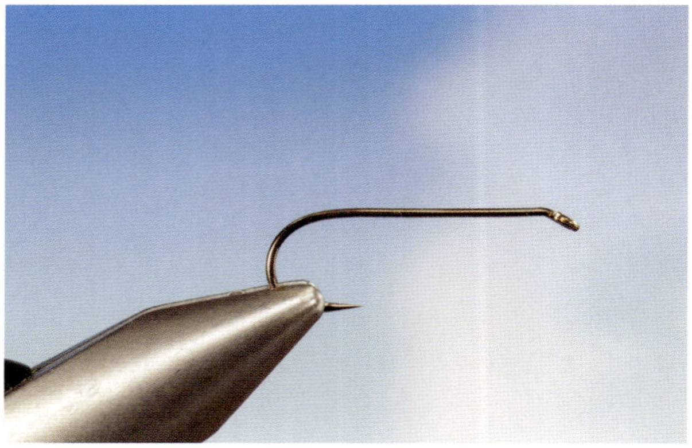

1 Squeeze the barb of the hook if you wish and place the hook in the vice.

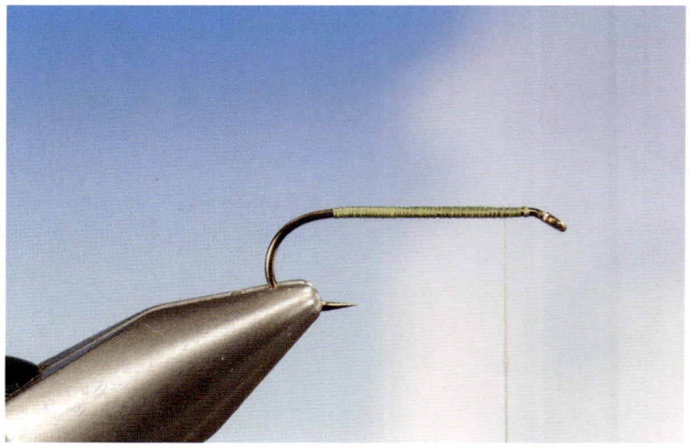

2 Cover the entire hook shank with closely spaced wraps of thread, starting from the eye.

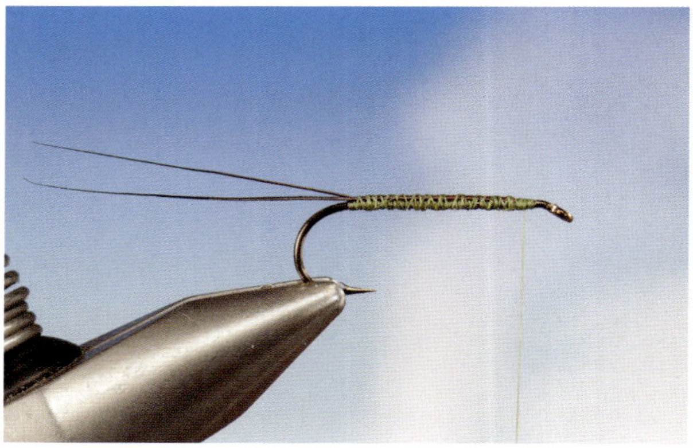

3 Tie in two thin peccary hair tips on top of the hook shank.

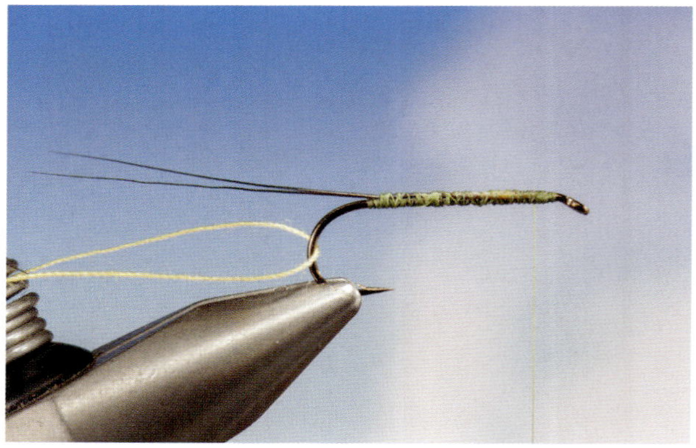

4 Put a piece of 3/0 thread around the bend and double it by putting the ends together.

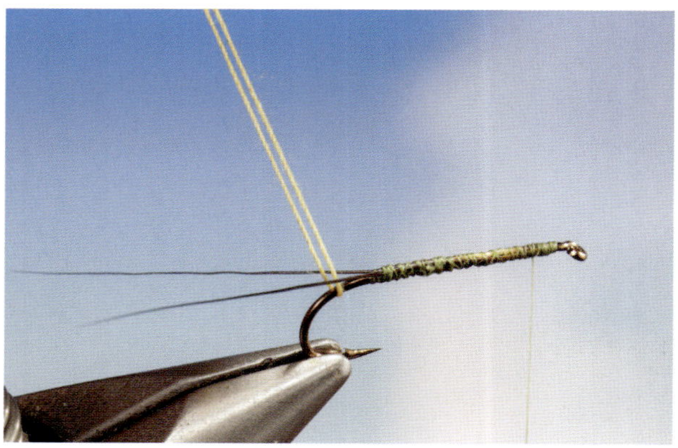

5 Split the tail in a perfect V shape by slowly lifting the thread strands between the tails.

6 Close up view on the forking trick.

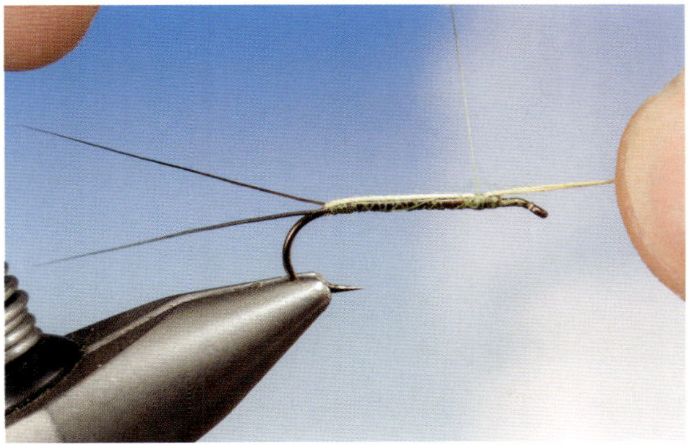

7 Lay the 3/0 thread perfectly on top of the hook shank and secure lightly.

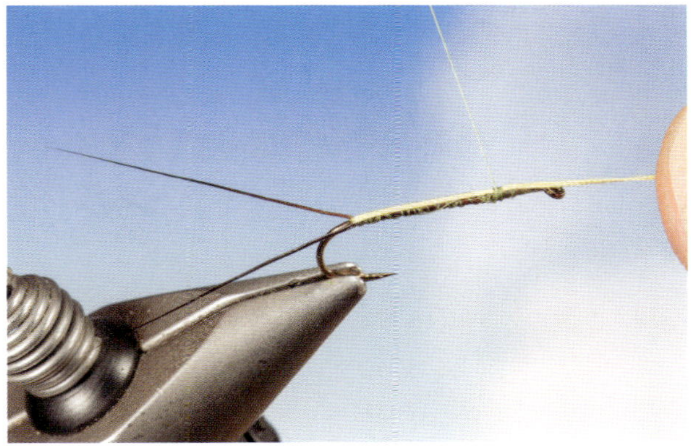

8 Then pull the thread strands towards the hook eye and pull it tight until the correct amount of split is achieved.

9 Now secure tightly and cut off waste then tie in the pure silk for the ribbing.

10 Tie in at least 5-7 feather fibres from a dyed wing or tail feather.

11 Wrap the body.

12 Secure the body well and cut off the redundant fibres.

13 Wrap your ribbing in the opposite direction, secure it with your tying thread and cut off waste.

14 Turn your hook upside down in the vice. Tie in about 5-7 fibres from the same wing feather and secure the fibres pointing backwards.

15 Turn the hook back to normal and tie in the hackle.

16 Wrap a nice collar with your hackle, but leave some space directly behind the eye for the wing. Secure the hackle and cut off the tip.

17 Turn the hook upside down again. Push the bottom hackle up a bit. Then pull the wing fibres over the hackles and secure them underneath the eye of the hook.

18 Clip off waste and varnish the fibres to give them extra protection and make the fly more durable.

19 Turn hook back to normal again.

20 See front view.

21 Tie in the wing over the hackles.

22 Secure well and tie off with your whip finisher.

23 The completed fly.

24 Front view.

25 Completed Green Glomma Dun tied with duck feather fibres.

The Scandinavia grayling usually turn black when they reach over 50cm (©Hein van Aar)

Two lovely Glomma Duns

One of the last Nalle Puhs that the late Lauri Sälevä tied, specially for this chapter

Chapter 11

Nalle Puh
The national fly of Finland

Since retiring I hold many small, private flytying workshops at my home. Most of these are held in peak flytying season, which is wintertime for me. It's a lot easier to invite people to my place than to travel around getting stuck in traffic jams and so on, because I really hate wasting time. My actual flyfishing workshops usually run from early spring to late autumn and are held on the riverbank, organised by lodges and resorts where I am staying. However, when people ask me for a flyfishing workshop in winter I mostly teach them how to set up a unique line, leader and tippet connection.

It's a system that's built up from almost knotless connections from backing, to fly line, to leader and finally the tippet. In separate workshops, I demonstrate cleaning and repairing flyfishing equipment, or share my fishing techniques, often with some additional help of video and computer presentations. At actual fishing locations, my most popular lessons are on how to read a river; such as where to expect to find fish, where the spawning routes the fish usually take are, where fish have their feeding lies and where you could expect a fish to grab your fly. I also cover how to recognise the food-bearing seams in surface currents, a few short lessons on insects as a background, and how to choose the best flies for a particular river or area.

During these visits, I have many people asking me to go through their fly boxes and examine the flies they have tied or copied. I judge their patterns quite objectively, especially in regard to the chosen materials and whether a fly will fish well or not. I also tell them how to improve the fly if it is possible and where, when and how to use the flies. I like to explain what are useful materials and what can be better left out. I always point out that superb fishing flies don't have to look like a piece of art. A simple demonstration is to let them see how my dry flies float in the bowl of water I always keep on my tying desk. Even after nearly fifty years of flytying, this water is still one of my most important 'tools' in my personal research of a pattern's merits. It's amazing how effective these little workshops really are, but also for me as a passionate flytyer and as a materials junky, it's quite interesting to see what materials people collect for flytying in their flytying boxes nowadays.

Simple thoughts

I know dozens of superb flies that don't look pretty at all; just have a look at the Rackelhanen, or my Leadhead, and I can give you many more examples of very rugged caddis flies I am crazy about. However, there's a big difference between just ugly patterns, and flies that are badly tied. Besides that, how many of us haven't had the experience that the more a fly picks up wear and tear while fishing, the more fish we catch with it? I can't help it, but I do love rugged and shaggy flies – and they catch lots of fish too.

You can work out a lot just by knowing and understanding what materials will do in the water, and how they behave in the tying process or even during casting. Having tied flies for so long I also notice that many people think and act very differently to the way I tie my flies and not many people seriously test their flies extensively before using them. I test all my new flies all the time. I meet a lot people at very exotic fishing locations with boxes full of flies that are completely useless, flies that have never been tested, or were tied with the wrong materials. **When making a fishing trip of a lifetime, you should never rely on just a few patterns and you should especially not tie boxes full of only one pattern.** While working at many lodges in spring and summertime I often saw this huge mistake.

Tying materials and why and how to test them

I always investigate new and possibly useful materials as soon as they come to my attention. I also do a quick appraisal as soon as I get my hands on anything new, especially using a material for the first time. When there is something that I don't like, I usually lose interest and stop using it right away. However, when I'm curious and finish a fly tied with new materials, and it passes my floatability and mobility tests, I will fish it as soon as I can. If possible, I test the fly in a swim tank that I keep outside in my little barn.

To understand the value of the best materials you need a dish of water to test your dry flies, terrestrials and floating emergers and a fly tester or swim tank is indispensable for assessing the mobility and action of your nymphs, flymphs, subsurface emergers, wet flies, streamers, saltwater flies or hairwings for Atlantic salmon. If I am satisfied with a new fly while fishing, I re-evaluate the whole tying process and write down the dressing and tying steps and fishing tips in my diaries and while I fish the fly I note improvements and changes I could make.

I recently inherited five big boxes of materials from an old flytyer who passed away and discovered a lot of unknown stuff in them. While going through these materials I noticed a large bag filled with some mysterious synthetic material with the highest floatability I had ever encountered since using polypropylene and similar products. I tried to push it on the bottom to sink it but it floated straight back to the surface. The nice thing was that it was available in several colours, but there was a

The last set of Puhs that Lauri Sälevä tied, specially for this chapter (© Antti Sorro)

Two very nice big CDC Puh flies tied by Juha Pusa who wrote Perhonsidonta together with Simo Lumme

big setback – the amount was limited to one bag only, making it hard for me to share. I love making my own dubbing so while testing this unknown synthetic, I blended it with other materials in my coffee grinder, and in this way I accidentally developed a new, extremely high floating and buggy-looking dubbing material. I often mix natural and synthetic materials and after yet more tests it turned out to be an absolutely magical dubbing for large and very large dry flies for trout, grayling and even Atlantic salmon.

During my third fishing trip for gigantic snakehead deep in the wilderness of Malaysia in 2018, people didn't believe me when I told them that all the Demon flies in my fly box were individually tested after they were tied. This is another reason why so many of my flies are so effective, because I know what my flies will do on, in, or below the surface before I actually fish them. It might look ridiculous, but I tell you it isn't. Using this strategy, I know exactly how the mobility of materials will work in my flies and whether the fly will do 100% of what I intended when tying it. Another reason I test large streamers before my trips is that after they have been wet I can shape them better after they dry again. This is also quite helpful when you use so many zonker strips as wing material, which I do. Bad quality strips often go hard when they dry out and can break, as I learned from testing zonker strips and because I use so many of them I mostly cut my own strips from good quality skins. Of course, mobility depends a lot on how you use your materials, or where you use the material in your tying steps. If you don't test this when you finish a new fly you will miss out on a very important aspect in your flytying. Experience will teach you which flies to test and which ones you can rely on without testing.

When I come across a new material my mind immediately starts thinking ahead. Is it durable enough to stand the power of casting? Will UV light have any influence on the material? Can it handle sharp trout teeth? Can I use it for dry flies, nymphs or streamers? Does it absorb water well or just the opposite? Is it good for creating added action in small streamers or large predator flies? Dozens of questions enter my mind which helps me enormously to understand the materials I'm using, which of them I like to work with and which ones I must add to my favourite materials after successful tests and plenty of fish catches.

I value my understanding of flytying materials and why I use certain materials for the flies I have developed over the years. Despite this, there is still so much more to learn and to test. I am not a person who scientifically explores and compares materials, but I do judge materials by results, by quality and by satisfaction during fishing. Catching success is the most important indicator that my choice of materials was right. Also, good mobility, floatability and durability in a fly are all solid indicators that I have used the correct and best materials in the fly that left my vice. Knowledge of materials has always been essential to me, but it really became indispensable after I stopped making fully dressed salmon flies and decided to concentrate on making fishing flies only. It was a very important decision because I had to make the choice between going full out for the art of flytying or to specialise more in making really good and fishable flies. Of course, making flies that are both artistic and good fishing flies is possible, but it takes a lot of time. I have always favoured patterns I could also tie quickly on a riverbank with a minimum of tools and effort.

When you talk about materials or write up the dressings for your patterns you must talk about materials that are available. This is why in the past I never talked about using Kapok in several of my dry flies. It's without doubt one of the best floating natural dubbing materials available in the world, but sadly my supply was so limited that I only had just enough for personal use. However, now that is available through Semperfli I openly can mention it.

Let's say a few words about my discussions and analyses of flies with the people who have visited me over the years, because I have made a few remarkable discoveries. Over the last twenty years, more and more flies are fully or partly constructed using glues and resins and this is clearly visible in many modern fly boxes. Of course, it's very handy using Super Glue to secure Dumbbell eyes onto your hook to prevent them for twisting, but glues and resins have their downside as well and have to be used wisely. I am not a big fan of glues and resins because I hold too many workshops with kids and they love to play with the UV lamps and glues and resins, which can so

Monster trout caught with Nalle Puh (© Antti Vanninen)

August 1962, Simo Lumme (right) and Lauri Sälevä near River Ivalo in northern Finland (© Esko Laurri)

easily go wrong. I also strongly believe that glues and resins may also damage some of the basic traditions and art and classic ways of flytying. I noticed this when I did workshops with groups of 8-12 year olds or even 12-16 year olds, and most of them just wanted to use resins on almost every fly they tied. Social media has had an enormous impact on the flytyers of the future. Maybe that's why of late I like to hold my classes and workshops in a more 'back-to-the-basics' manner.

When I learned flytying it took several attempts because I simply had no clue and flytying books were rare and the Internet didn't exist. I am from a generation where silk was used as tying thread and floss, and seal and hare were extremely popular as body material. I was just extremely lucky that I ran into the world of poly yarn and extra fine poly dubbings in the early 1980s through which I was able to find my own way. I am all for innovation, but not all new materials hold my favour, or are better than the materials and methods used in the past. I am concerned that in a few years' time several tyers will just glue their flies together instead of tying them.

In addition, certain synthetics can be extremely dangerous when Super Glue is added to them and may emit very toxic vapour that you really don't want to get into your lungs and eyes. Accidents with Super Glue are also not uncommon. Super Glue can also destroy the properties of certain material or entire flies, when people use it to secure wings and parachute hackles. I have seen boxes full with commercially tied Klinkhåmers that were totally useless, as they had lost their floatability because the wing base and part of the wing had been hardened with Super Glue instead of varnish. It totally destroyed the all-important absorbing ability of the poly yarn. It made the use of floatant impossible which is a crucial consideration for flies with poly yarn wings. Super Glue has good uses too and I have saved several flyfishers' trips by using Super Glue to close their wounds as I was taught to do on the battlefield in the army. It works above all expectations and leaves no nasty looking scars.

Not long ago I had a friend complain that all his dry flies were sinking and when I had a closer look I saw the problem right away. The flies were unbelievably beautiful, but the bodies were covered with a thick layer of resin hardened by an UV lamp to create an extra good-looking body. The bodies indeed were awesome, but the resin made the flies so heavy that they sank rather quickly. Also, most floatants don't work on hardened resins at all. Although resins can create amazingly beautiful flies, you still have to know and understand when, why and how much you need to apply.

This all led me to the thinking behind fly patterns and why people should really take note of the intentions, methods and advice from the developer of any good and successful fishing pattern. Lots of people change the original dressings of flies to materials that they prefer, or just replace it because it's simply a lot cheaper. Another reason can be that the materials are hard to find or unavailable. Please note that not all changes are bad and I know a lot of flies that were greatly improved by changing the original materials. In the past, most flies were tied with materials that were easily obtained. A perfect example is the North Country series of soft hackle patterns where the hackles were from birds common in those days. In Newfoundland and Labrador, all the traditional wet salmon flies had wings of caribou or moose hair because it was so easy to get. It worked well because the Atlantic salmon in that part of the world really love to grab wet flies that skate on the surface, often with the angler employing an extra hitch. I never had any luck with those flies in Scandinavia, but by swapping the stiffer hair on the original flies with much softer fitch or squirrel hair, I did a lot better using my hairwing patterns subsurface than any of the original flies ever did.

The national fishing fly of Finland

The materials that fly designers use are chosen for extremely important reasons. That's why I am always curious about the strategies behind fly patterns. With flies from the past we often have serious problems because some of the old materials are now extremely hard to find, or have became illegal to possess, or even worse are extinct. This leads me to a fly that is almost impossible to make with the original materials. People who live in Scandinavia, or visitors who fish there a lot are probably familiar with the pattern. I am talking here about the rugged-looking fly with the amazing name of 'Nalle Puh'.

Mr Simo Lumme (1928-1998) was a Finnish architect who had enormous passion for flytying and flyfishing. He appreciated and admired

In eastern Canada, Atlantic salmon will regularly take a Nalle Puh

This is how my perfect Nalle Puh water is looking in Finland (© Antti Vanninen)

Awesome rapid for a large Nalle Puh fly in Canada

This wonderful and rare picture of Simo was taken in 1994 by Barry Ord Clarke while he was visiting Simo
(© Barry Ord Clarke)

The original picture of the Nalle Puh that Simo tied for the book Perhonsidonta which he wrote with Juha Pusa
(© Simo Lumme and Juha Pusa)

The only flies I could find that were tied by Simo

Simo Lumme at the yard of the log house in Kärekeniemi 5 July 1967 (© Esko Lauri)

G E M Skues' way of thinking and how he tied his flies, and the way of designing caricatures and not total imitations. It is something else Simo and I have in common because I always had an enormous respect for Skues living in a world of so many dry fly purists and the controversy that it caused with the chalkstream dry fly doctrine developed by Frederic M Halford and others. I also love people going their own way.

The Nalle Puh is a perfect example of a caricature type of fly, in which the main idea is to imitate the fluttering adult and egg laying caddis present on fast flowing rivers, and especially rapids. Simo is not just the inventor of the Nalle Puh but also the man who helped define a country's love for flyfishing. I was in Finland many times in the 1970s and 1980s and I really feel it a huge pity that I never met him.

The Nalle Puh is without doubt Simo's most famous dry fly pattern. When he developed the Nalle Puh, a good name came quickly to his mind. It is the well-known Finnish name for the fictional character Winnie-the-Pooh created by English author A A Milne in 1926. So the Nalle Puh became famous among flyfishers all over Scandinavia and with foreigners who frequently fished in Finland, like me.

Today the Nalle Puh is regarded as the National Fly of Finland and is the best-known Finnish fishing fly around the world. Whether he chose the name because the Nalle Puh has a sort of 'two ears' like Winnie the Pooh, or because he actually used bear hair for the wings, is still unclear, but perhaps it was for both reasons, which makes sense to me. Nalle is a nickname for a bear, more a Teddy bear to be precise. From a close Finnish friend I heard that after his fly turned out to be so effective, Simo became really worried they would kill bears to get hair for the wing, which was definitely not his intention.

Why I put this extraordinary pattern in my book

Most of my flytying is related to the countless fishing trips I have done all over Scandinavia since 1971 and so when I had my first plan to

Simo Lumme with his pipe, 5 July 1967, River Nuortti in the eastern part of northern Finland (© Esko Lauri)

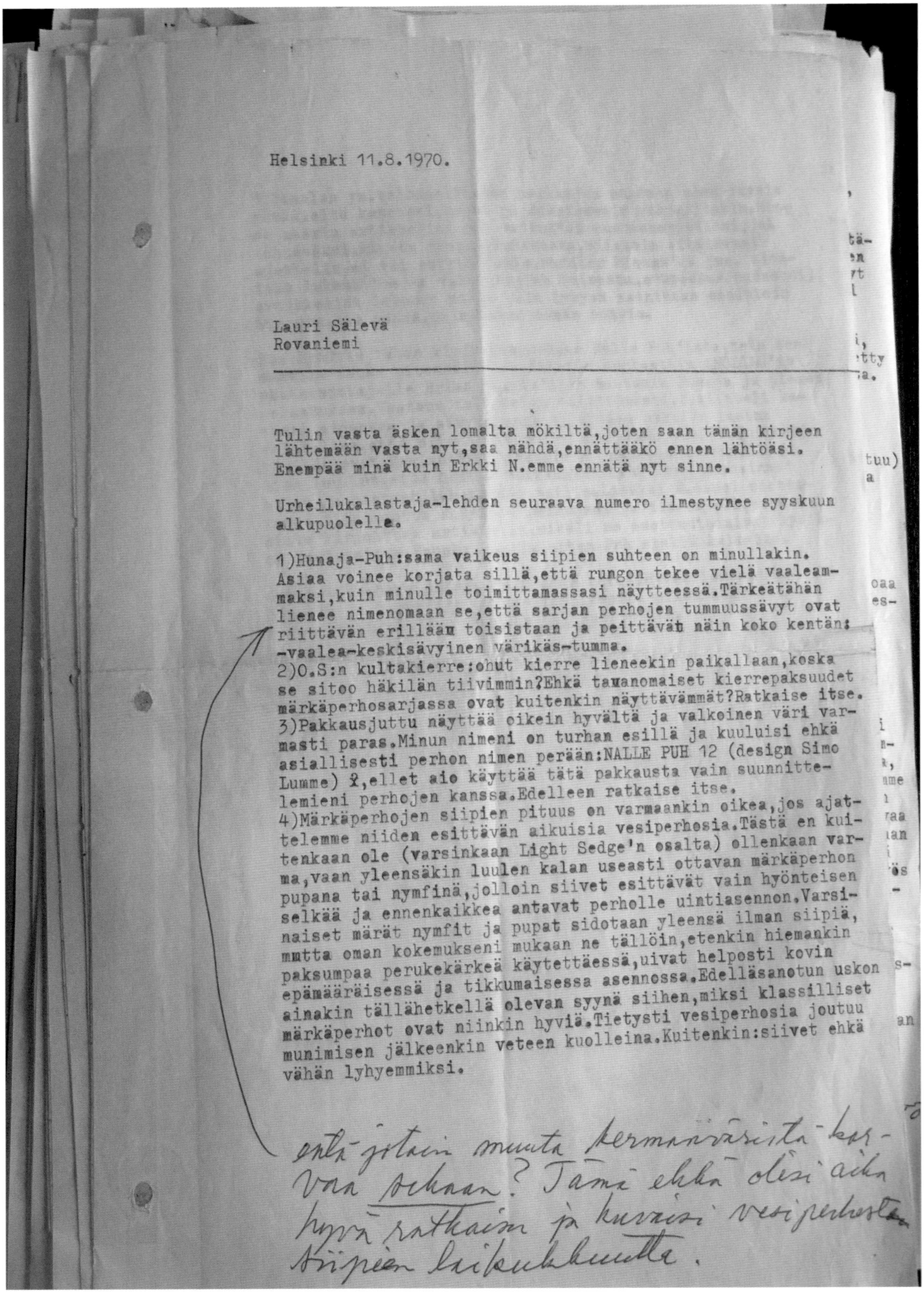

Helsinki 11.8.1970.

Lauri Sälevä
Rovaniemi

Tulin vasta äsken lomalta mökiltä,joten saan tämän kirjeen lähtemään vasta nyt,saa nähdä,ennättääkö ennen lähtöäsi. Enempää minä kuin Erkki N.emme ennätä nyt sinne.

Urheilukalastaja-lehden seuraava numero ilmestynee syyskuun alkupuolella.

1)Hunaja-Puh:sama vaikeus siipien suhteen on minullakin. Asiaa voinee korjata sillä,että rungon tekee vielä vaaleammaksi,kuin minulle toimittamassasi näytteessä.Tärkeätähän lienee nimenomaan se,että sarjan perhojen tummuussävyt ovat riittävän erillään toisistaan ja peittävät näin koko kentän: -vaalea-keskisävyinen värikäs-tumma.
2)O.S:n kultakierre:ohut kierre lieneekin paikallaan,koska se sitoo häkilän tiivimmin?Ehkä tavanomaiset kierrepaksuudet märkäperhosarjassa ovat kuitenkin näyttävämmät?Ratkaise itse.
3)Pakkausjuttu näyttää oikein hyvältä ja valkoinen väri varmasti paras.Minun nimeni on turhan esillä ja kuuluisi ehkä asiallisesti perhon nimen perään:NALLE PUH 12 (design Simo Lumme) ?,ellet aio käyttää tätä pakkausta vain suunnittelemieni perhojen kanssa.Edelleen ratkaise itse.
4)Märkäperhojen siipien pituus on varmaankin oikea,jos ajattelemme niiden esittävän aikuisia vesiperhosia.Tästä en kuitenkaan ole (varsinkaan Light Sedge'n osalta) ollenkaan varma,vaan yleensäkin luulen kalan useasti ottavan märkäperhon pupana tai nymfinä,jolloin siivet esittävät vain hyönteisen selkää ja ennenkaikkea antavat perholle uintiasennon.Varsinaiset märät nymfit ja pupat sidotaan yleensä ilman siipiä, mutta oman kokemukseni mukaan ne tällöin,etenkin hiemankin paksumpaa perukekärkeä käytettäessä,uivat helposti kovin epämääräisessä ja tikkumaisessa asennossa.Edelläsanotun uskon ainakin tällähetkellä olevan syynä siihen,miksi klassilliset märkäperhot ovat niinkin hyviä.Tietysti vesiperhosia joutuu munimisen jälkeenkin veteen kuolleina.Kuitenkin:siivet ehkä vähän lyhyemmiksi.

entä jotain muuta kermanväristä -korvan sekaan? Tämä ehkä olisi aika hyvä ratkaisu ja kuvaisi vesiperhosten siipien liikehdintää.

The box with all the letters, sketches, articles and drawings sent from Simo to Lauri that Ari showed me at the Fish Library

write a book, I knew right away I would include some Scandinavian patterns. The idea to include a chapter on the Nalle Puh in this book came after I had seen some of Simo's original work and was completely in awe of his incredible and detailed personal notes and letters. Most of them provided information that is absolutely original and crucial to make exact copies of his flies. Beside that I had never seen such exhaustive personal notes on a series of flies with added drawings and real materials included that are more than fifty years old and still in good shape and colour tones. I regard myself very fortunate to have so many good friends in Finland nowadays and with their help I was able to get some of Simo's personal work and was able to discuss Simo's thoughts and ways of thinking. When I received the parcel in my mailbox and looked at it I jumped for joy. When I then got official permission from the late Lauri Sälevä, who owned all the copyrights after Simo's death, I knew right away that I had something extraordinary to show and share with the world. Lauri and Simo were very close friends.

The first idea was to use everything in a story I had planned to write about the Nalle Puh, but when I dug deeper into the history of the Nalle Puh I found the information was worthy of a future book. Antti Sorro and Matti Rantanen were my main contacts who arranged everything for me. Sadly Lauri died in 2012. Antti Vanninen and Juhani Kyllönen provided wonderful help as well. Antti is my oldest Finnish flyfishing friend, a friendship of more than three decades. Antti wrote a beautiful story for the *Dutch Fly Fishing Magazine* in the early 1990s about flyfishing in the pristine wilderness of Kuhmo that I translated for our readers into Dutch.

While writing here for this chapter he, as usual, did a lot for me behind the scenes. Juhani had become a close friend during the April Orange events in Kuhmo in 2018 and 2019, when we spent a lot of time together and he did most of my translating because the Finnish language is not easy. With the English translations I could

do much better research on the Nalle Puh. Juhani had developed an enormous passion for the Kuhmo area of Finland and all the rapids and its fish. His conservation work there was close to my heart and was extremely important to Kuhmo and for the flyfishing in the entire region. Anne Hyytiäinen, another good friend, did some translations as well, but more importantly, she acquired a beautiful patch of brown bear fur for me with all the legal papers. This made it possible for me to tie original Nalle Puh flies again after a long lapse. A few other people who helped me a lot in drafting this chapter were Juha Pusa, Simo Yli-Lonttinen, Anni Yli-Lonttinen, Kirsi Piirainen and Ariko Miettinen. In my opinion Ariko is one of the best flytyers in Finland and he currently ties the best-looking Nalle Puh flies. That is why I am so happy he tied a complete series especially for this chapter. Every time I was stuck in my research, I could rely on the help of Juha Pusa to pick it up again. Juha also hooked me up with Aki Rinne who was a great help on the background of the sedge species which are so important in the history of the Nalle Puh. Without the help of all these people I would never have been able to write this chapter or stay so highly motivated when inspiration was lost. All I can say is thanks a million, guys, for your amazing help.

The Finnish Fish Library

In 2023 I uncovered new information about the Nalle Puh at a private Finnish Fish Library (Suomen Kalakirjasto) in the wilderness of southern Lapland on 26 June. The Library was close to the village of Muonio and the trip was organised by my good friend Elias Teriö. We agreed that I would look around for a day and do a tying demonstration in the library in the afternoon for the owner and guests at the library.

Ari Savikko is the big man behind the library collection which represents a thirty-year journey in which every item collected has its own story. The immense collection is a chronological record of fishing in Finland from 1730. The collection is Ari's desire to preserve an important part of Finnish culture and fishing history for future generations. When his collection didn't fit in his house anymore he decided to build a brand-new library. The library has become a way of life for Ari. Ari designed the Library himself, with help from his brother-in-law and built it on land owned by his father-in-law. The library was opened on 25 August 2018.

To be honest, the Library was more like a museum to me and I easily could have stayed there for a whole week. There were approximately 53,300 books or writings on the database, including all professional magazine articles published prior to the 1940s. The database has subsequently been updated to include all articles published before the 1950s. In addition, the collection also includes nearly every Finnish fishing book and magazine published to date. These are also continually updated to the database. You can find some English books as well.

After talking with Ari about my interest and passion for Simo Lumme's work he showed me some files that were not registered during my visit. But I was able to see a lot more from Simo Lumme's original work that had been donated to the Library after Lauri Sälevä had passed away. I dug for another 30-40 minutes into a goldmine

Finnish Fish library www.suomenkalakirjasto.fi

Top: Ari's biggest grayling was 2cm bigger than mine in 1981
Above: Talking seriously with Ari about Simo Lumme's work
Left: one of the book corridors packed with amazing fishing and flyfishing books
Bottom: Ari found the box with all Simo's files and artwork

Juha Pusa who wrote, together with Simo Lumme, the book Perhonsidonta (© Juha Pusa)

of flyfishing history. In moments like these I regret that I can't speak the Finnish language. Because of my respect for Ari and all the work he has done to create the Library, I left a dozen flies behind as a gift for the collection.

Which Puh came first?

During my research a few people told me that the Dark Puh (imitation of the *Arctopsyche ladogensis*) was developed before the Nalle Puh, and that it was designed for the north, while the Nalle Puh was developed a little later to imitate the *Potamophylax latipennis*, which is more abundant in the central part of Finland. Juha Pusa, who wrote the book *Perhonsidonta* (Fly Tying) together with Simo Lumme, thinks this indeed might be the case. Simo and Juha never talked about which of the Puh patterns was actually developed first so we really don't know which one left Simo's vice first and obviously we can't ask Simo anymore. In his first article in '*Urheilukalastuslehti*' (Sportfishing magazine no 3/1969), there is a comment that the Dark Puh was developed for late evening and night fishing. He also states clearly that at the Nuorttijoki river he did not have such a large dark Caddisfly. After the publication of Simo's second article (*New Caddis Imitation*) in 1970, things got a little confusing for me but also for a few of my Finnish friends who were familiar with the 1969 story.

For myself it was an easy decision because after I read an excellent translation of Simo's 1970 article and got some excellent advice from Juha Pusa, I decided to take the 1970 story as my starting point, just to prevent confusion.

What Juha also told me was that Simo tied the Dark Puh on larger hooks than the original Nalle Puh or Honey Puh, often a size 4 hook, and most probably also on long shank down-eye streamer hooks in sizes 6-8. The more I started to know about Simo, the more my respect for him grew and the more I felt in common with my flies and style of fishing. Simo did not talk much about these large over-sized Puhs in public and I can understand why because I was equally reticent about my huge oversized Klinkhåmers and Leadheads, as not many people believed in them.

In the beginning, Simo used dark squirrel tail as a wing, which unlike the normal *Trichoptera* imitations he tied, were vertical, so the fly would give the impression of a female egg-laying caddis moving over turbulent water. The fly got him some fish, but the wings looked dead after the use of liquid floatants such as Permaflote, or when the fly got wet. Then the wings quickly formed a thin string-like bunch. The wing problem remained unsolved until he visited Lapland again and got a few pieces of different coloured bear fur and was eventually satisfied with the result. Simo mentioned in his article that the curly bear hair was partially translucent so it broke up the sunlight into different colours and shades thus increasing the impression of life and movement.

The Nalle Puh and some crucial background information

Simo's Nalle Puh was not born by chance, but after a long period of intensive observation and research. Simo wanted to design a dry fly for fast flowing rivers and countless rapids in Finland. This is another important reason why I feel a little sad that I never met him. I too love big and highly visible fast water flies, so I am sure we were of the same mind.

Simo found the inspiration for his fly in Finnish Lapland on the River Nuortti in the first week of July, 1968. He told his friends that he wanted to find a usable imitation of the colourful and larger cinnamon sedge, *Potamophylax latipennis*. The species was known in the north as *Stenophylax stellanus*. Finnish novelist and flyfisher Juhani Aho described it as having a brown-winged, yellow body that was big and fat. Simo Lumme had his Orange Sedge imitation, which had turned out to be extremely successful using a body blended from orange, yellow and light brown colours.

He took this blend into his new bear-winged fly but I think his most notable breakthrough was his discovery that wings made from the fur of a young bear were more effective. He cut the hairs close to the skin and tied in the wings at their tips, actually upside down, which at that time was a very novel way to tie a fly. With this new and unique flytying step he was trying to mimic the female egg-laying caddis with fluttering wings. At many shows, fairs, video clips and step-by-step tying explanations in magazines and online, I have watched people tying the Nalle Puh and

The book Perhonsidonta (Flytying) which Simo wrote with Juha Pusa

most tie the wing with the hair tips on top, which is of course one of the biggest mistakes people can make when trying to copy the Nalle Puh. The only flytyer I ever saw using a similar technique was Branko Gašparin of Slovenia who used a reverse sedge wing technique with fox hair for making special sedge wings in a few series of the sedge patterns he created.

In almost everything I know about Simo today I can sense his fascination and passion for caddisflies. After reading his letters I know how he must have examined the stomach of the fish he caught and even drawn sketches of them on the riverbank. While fishing in Scandinavia so often, I rapidly learned that caddisflies dominate most of the waters in Finland, and while so abundant, it also meant they are an important food source for trout, grayling and several other species.

As far I know there are well over 200 caddis species in Finland and that makes caddisfly imitations indispensable for flyfishers in the north. A German University even claims (on the Internet) that there are somewhere between 6,000-7,000 caddis species in Scandinavia and the Baltic states alone. What really makes sedges more interesting and exciting is the fact that there are some excellent books about caddisflies in Finland and in the English language too, like *Trichoptera Larvae of Finland* by Aki Rinne and Peter Wiberg-Larsen.

The natural insects described by Aki Rinne

Arctopsyche ladogensis

Development takes two years. Larvae are predators, feeding on small water invertebrates and on detritus. Wing length in males is between 10.2-13.6mm and for the females between 13.2-18.1mm. Wings are dark grey, sometimes with brownish spots. Body and legs are dark in females while the abdomen is a distinctly dull olive-greenish colour. Neck is distinctly light green and the abdominal lateral line greyish-green. Females lay their eggs on the water surface. Pupae are very effective imitations, as are emergers. Males are not important for fishermen at adult stage. Species is abundant and numerous all over northern Scandinavia, being distinctly bigger than all its close families (*Hydropsychidae*). In Lapland species emerge from mid June and is over by mid July lasting locally only two weeks. Imitation: Dark Puh.

Potamophylax spp.

Development takes one year. At larval stage all species are active and mobile all months

Potamophylax latipennis – Imitation: Nalle Puh (© Aki Rinne)

Arctopsyche ladogensis – Imitation: Dark Puh (© Aki Rinne)

of the year. Larvae are shredders, feeding on decayed plants and leaves of trees. In Europe and Scandinavia living species all have roughly the same appearance. The length of front wing varies from 15-24mm. At an adult stage all the species looks practically the same. Pupae are predominantly orange-coloured, head, legs and wing pads being a slightly darker shade of light brown. Before metamorphosis, larvae move nearer shores, where they anchor their cases beneath stones and submerged logs etc. All pupae swim or crawl on solid objects at night time, where they cast off their pupal skin (exuviate) in a few minutes. Males are unimportant to flyfishers. Females lay their eggs on moist splashy stones, sometimes covered by water plants (*Fontinalis spp.*). All species of the genus are active at dusk.

Habitat in Finland

P. latipennis: Abundant and numerous in rapids of medium and large temperate rivers. Locally in lakes of Lapland. Not found in springs and spring brooks. Imitation: Nalle Puh.

P. rotundipennis: Abundant and numerous in temperate and warm ditches and various sized rivers. Not found in springs and spring brooks. Not found in cold waters of Lapland. Imitation: Honey Puh.

P. cingulatus: In southern provinces locally in groundwater-fed brooks, in north common in brooks and tributaries of large rivers. In Lapland, common and abundant in brooks and rivers. Missing from lakes. Imitation: Honey Puh.

P. nigricornis: In south locally, living only in groundwater-fed spring brooks. In Arctic mountain areas common in all cold running waters. Typical Arctic relict which inhabits glacier and snow melt waters. Imitation: Nalle Puh (in spring creeks), Honey Puh in Lapland.

Potamophylax cingulatus - Imitation: Honey Puh
(© Aki Rinne)

Dark Puh tied by Ariko Miettinen

Original Nalle Puh tied by Ariko Miettinen

The final Nalle Puh

The perfect Nalle Puh fly was developed at Simo Lumme's home in Helsinki in the autumn of 1968. It was so outstanding that the Nalle Puh rapidly gathered a heap of friends among flyfishers. At that time the Nalle Puh was a distinctive fly; it was a better fly than so many others, or even the only fly to fish successfully on rapids, which are the waters flyfishers have always been so keen to fish in Finland. So there was a huge demand to have the Nalle Puh on sale. With Simo Lumme's busy job he had no time to produce Nalle Puhs in great numbers, but a solution was quickly found when he befriended Lauri Sälevä in Rovaniemi. Lauri was a construction site foreman and they met for the first time at a construction project. Beside his job he tied flies for sale so a helping hand was near, despite 700km separating them.

Lauri was the man behind the original Nalle Puh flies on the market. As a perfectionist, Simo Lumme demanded only the best quality commercially tied flies. Consequently, there were amazingly good tying instructions available that Simo had specially made and drawn for Lauri.

In his 1970 article Simo wrote: *'Those fly fishers who'd like to try the previously explained sedge/caddis imitations but who don't tie their own flies could be interested that I have delivered original detailed instruction with component samples to my good friend and fishing buddy and famous fly tier in the north Lauri Sälevä'.*

I was so happy to have finally found two Nalle Puhs tied by Simo himself, and I feel very fortunate and proud to have the last three Nalle Puhs tied as a complete series by Lauri. He gave them to me especially to use in this book.

Simo's private letters to Lauri

The private letters written by Simo and sent to Lauri are a treasure trove to me. They are filled with essential information and tips based on his very personal thoughts about tying the Nalle Puh and its variations. I am convinced that Simo put all this exclusive information on paper to be absolutely sure that Lauri would tie his flies exactly as he had it in mind. In his first letter to Lauri, Simo explains in handwritten notes how he is working on final instructions for him in which he will explain and describe all the colours and shades for tying the perfect Nalle Puh series.

The letter contains information about the background history of his pattern. A few other letters were typed and accompanied by handwritten notes and additional fly or insect drawings to explain his thoughts and techniques more clearly. Other letters he wrote entirely by hand but with the actual materials, fly parts and flies stuck on the letter with transparent cello tape. I even have a letter about flies and fly parts that Lauri had sent to Simo and which Simo returned with his notes and comments on improvements. The most valuable letter by far is the one in which Simo put together the dubbing mixes of seal's and hare's furs for the bodies of the Nalle Puh, Honey Puh and Dark Puh. In that same letter he also addressed the wing and made drawings on how and where exactly to tie in the brown bear hair on the hook shank. Again, everything is stuck to the letter with Sellotape. Many of Simo's drawings were done with an ordinary ballpoint pen.

Another valuable letter that Simo sent to Lauri is the one with his ballpoint drawings of the tying steps of the Nalle Puh where he describes, using a typewriter, the ten tying steps. The six tying steps for the book that he wrote with Juha Pusa were done with pencil and proved to me that he was indeed also a great artist.

First Nalle Puh letter from Simo to Lauri

In this five-page letter, which was typed and supplemented with handwritten comments, Simo referred to his article in *Urheilukalastus*. He mentioned that he got the impression that caddisflies played a significant part in fishing rapids in Finland. He also explained that in certain

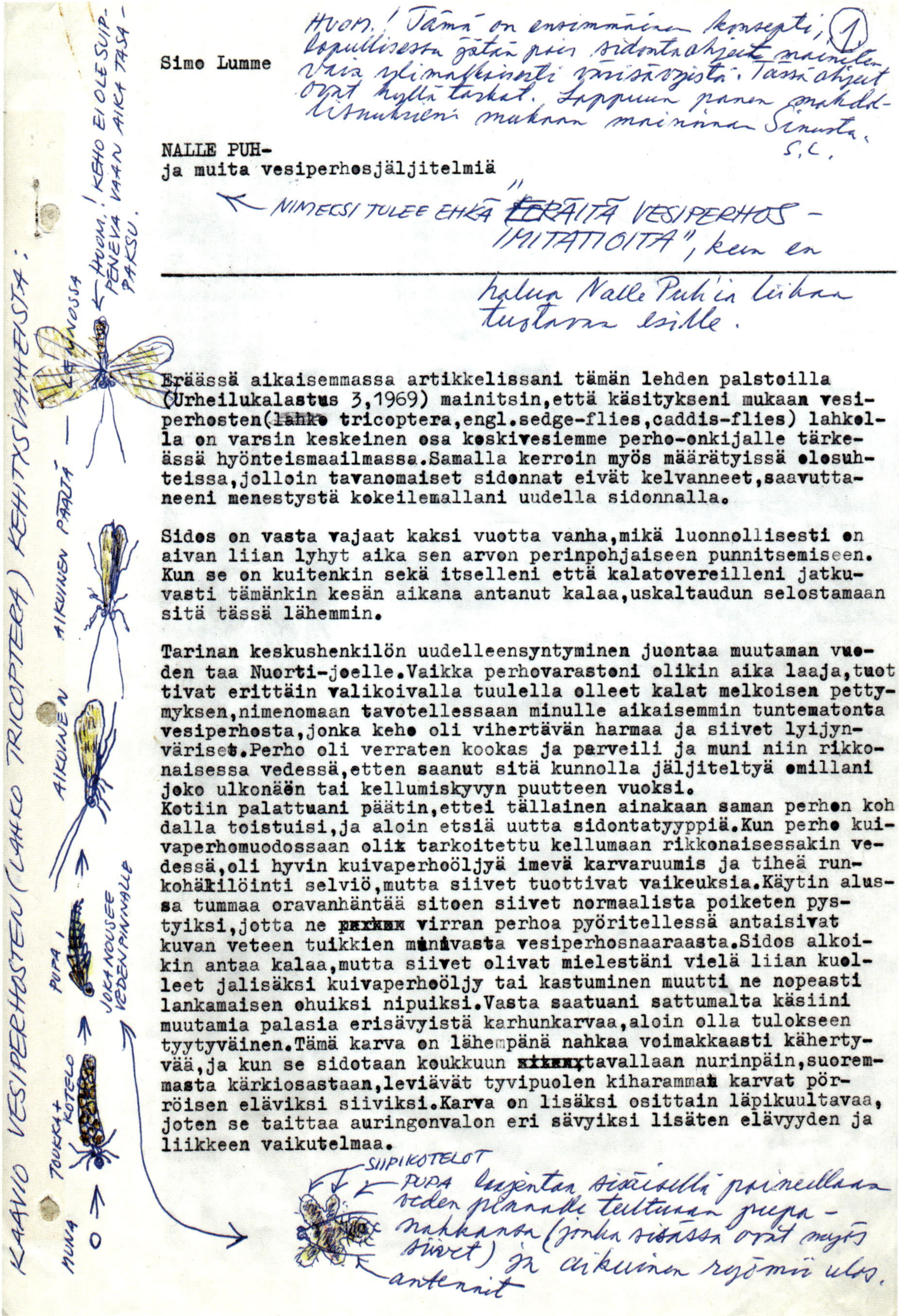

Simo Lumme

NALLE PUH-
ja muita vesiperhosjäljitelmiä

Eräässä aikaisemmassa artikkelissani tämän lehden palstoilla (Urheilukalastus 3,1969) mainitsin,että käsitykseni mukaan vesiperhosten(lahko tricoptera,engl.sedge-flies,caddis-flies) lahkolla on varsin keskeinen osa koskivesiemme perho-onkijalle tärkeässä hyönteismaailmassa.Samalla kerroin myös määrätyissä olosuhteissa,jolloin tavanomaiset sidonnat eivät kelvanneet,saavuttaneeni menestystä kokeilemallani uudella sidonnalla.

Sidos on vasta vajaat kaksi vuotta vanha,mikä luonnollisesti on aivan liian lyhyt aika sen arvon perinpohjaiseen punnitsemiseen. Kun se on kuitenkin sekä itselleni että kalatovereilleni jatkuvasti tämänkin kesän aikana antanut kalaa,uskaltaudun selostamaan sitä tässä lähemmin.

Tarinan keskushenkilön uudelleensyntyminen juontaa muutaman vuoden taa Nuorti-joelle.Vaikka perhovarastoni olikin aika laaja,tuottivat erittäin valikoivalla tuulella olleet kalat melkoisen pettymyksen,nimenomaan tavotellessaan minulle aikaisemmin tuntematonta vesiperhosta,jonka keho oli vihertävän harmaa ja siivet lyijynväriset.Perho oli verraten kookas ja parveili ja muni niin rikkonaisessa vedessä,etten saanut sitä kunnolla jäljiteltyä omillani joko ulkonäön tai kellumiskyvyn puutteen vuoksi.
Kotiin palattuani päätin,ettei tällainen ainakaan saman perhon kohdalla toistuisi,ja aloin etsiä uutta sidontatyyppiä.Kun perho kuivaperhomuodossaan olik tarkoitettu kellumaan rikkonaisessakin vedessä,oli hyvin kuivaperhoöljyä imevä karvaruumis ja tiheä runkohäkilöinti selviö,mutta siivet tuottivat vaikeuksia.Käytin alussa tummaa oravanhäntää sitoen siivet normaalista poiketen pystyiksi,jotta ne virran perhoa pyöritellessä antaisivat kuvan veteen tuikkien vasta vesiperhosnaaraasta.Sidos alkoikin antaa kalaa,mutta siivet olivat mielestäni vielä liian kuolleet jalisäksi kuivaperhoöljy tai kastuminen muutti ne nopeasti lankamaisen ohuiksi nipuiksi.Vasta saatuani sattumalta käsiini muutamia palasia erisävyistä karhunkarvaa,aloin olla tulokseen tyytyväinen.Tämä karva on lähempänä nahkaa voimakkaasti käherty-vää,ja kun se sidotaan koukkuun tavallaan nurinpäin,suoremmasta kärkiosastaan,leviävät tyvipuolen kiharammat karvat pörröisen eläviksi siiviksi.Karva on lisäksi osittain läpikuultavaa, joten se taittaa auringonvalon eri sävyiksi lisäten elävyyden ja liikkeen vaikutelmaa.

The oldest letter that I found with comments and even drawings of a caddis hatch (© Simo Lumme)

conditions when the fish did not accept regular flies he succeeded with flies he had tied using a completely new tying method. After he had used the fly for two years he concluded that it was too short a time to draw thorough conclusions, but because of the enormous success that he and his friends had achieved, he felt he could now explain a lot of things in far more detail. Let's share with you what Juhani Kyllönen translated for me from this personal letter because it is invaluable to listen to Simo's own words.

'The rebirth of the head role player in this story happened a few years ago at river Nuortti. I have quite a selection of flies in my box. Still the fish were very selective and I was unsatisfied with my results. The fish were feeding on one special fly, which was previously unknown to me. The fly's body was greenish grey in colour and the wings were the colour of lead. The fly was relatively sizeable and was swarming and laying eggs in water which was not calm. I wasn't able to imitate the fly – the look and the floating ability were not there.

After I had returned home I decided that would not happen again. Not with that fly anyway. So I started to find a new tying method. Since the fly in its dry fly form was supposed to float even in fast, 'broken' water the body had to be furry and absorb a lot of floatant. It also had to have very thick hackles on the body. But the wings caused me trouble. In the beginning I used dark squirrel tail and tied them unconventionally upwards so they'd look like wings of an egg-laying female fly in flowing, whirling water. That fly got me some fish, but I felt the wings were too 'dead' and the floatant and water eventually turned the wings to string-like bundles. But when I got pieces of bear fur in different colours I started to be satisfied with the results. Close to the skin the bear fur is very curly and when it is tied on a hook 'upside down' i.e. from the tip end of the hair, the curly bottom end of the hair spreads out making fluffy and alive-looking wings. The hair is also partially transparent so it refracts sunlight in different colours and shades increasing the sense of life and motion'.

My first Nalle Puhs

I have known the Nalle Puh series for a long time. I even fished extensively with the original and also with my own versions in the wild and rugged rivers of Scandinavia from the mid 1970s until the end of the 1990s. After that I kept on fishing Nalle Puhs, but with wings from poly yarn. I also regularly used the Puh series in Canada, but always in rivers with a strong current or rapids and especially for brook trout in eastern Canada. Really large Puhs did extremely well for grayling in the Yukon Territory under the same conditions. I tried them very successfully in Mongolia and Alaska as well. I call my synthetic

Original Puh series as tied by Ariko Miettinen (right one with poly yarn wing)

uusi vesiperhosjäljitelmä

Nalle Puh

Simo Lumme

Aikaisemmassa artikkelissani tämän lehden palstoilla (Urheilukalastus 3, 1969) mainitsin, että vesiperhosten lahkolla (tricoptera, engl. sedge-flies, caddisflies) on käsitykseni mukaan koskivesiemme perho-onkijalle varsin keskeinen osa. Samalla kerroin myös määrätyissä olosuhteissa, jolloin tavanomaiset sidonnat eivät tuottaneet tulosta, saavuttaneeni menestystä kokeilemallani uudella kuivaperhotyypillä.

Tämä sidos on vasta parisen vuotta vanha, mikä luonnollisesti on aivan liian lyhyt aika sen arvon perusteelliseen punnitsemiseen. Kun se on kuitenkin sekä itselleni että kalastustovereilleni tämänkin vuoden aikana jatkuvasti antanut kalaa, rohkenen selostaa sitä tässä lähemmin.

Perhon kehittelyn alku juontaa muutaman vuoden taa Nuorttijoelle.

Vaikka mukanani ollut perhovarasto olikin aika laaja tuottivat erittäin valikoivalla tuulella olleet kalat melkoisen pettymyksen, nimenomaan tavoitellessaan minulle aikaisemmin tuntematonta tummaa vesiperhosta, joka oli niin kookas ja parveili niin rikkonaisen veden yllä, etten saanut sitä omillani kunnolla jäljiteltyä ulkonäön ja kellumiskyvyn suhteen.

Kotiin palattuani päätin, ettei tällainen toistuisi ainakaan saman perhon kyseessä ollen. Kun perhon tuli kellua rikkonaisessakin vedessä, oli hyvin kuivaperhoöljyä imevä karvarunko ja runkohäkilöinti selviö, mutta siivet tuottivat vaikeuksia. Käytin alussa tummaa oravanhäntää sitoen siivet normaaleista tricopteraimitatioista poiketen pystyyn, jotta ne virran perhoa pyöritellessä antaisivat kuvan veteen tuikkien munivasta vesiperhosnaaraasta. Sidos alkoikin antaa kalaa, mutta siivet olivat mielestäni vielä liian kuolleen näköiset, ja lisäksi kuivaperhoöljy tai kastuminen muutti ne nopeasti lankamaisen ohuiksi kimpuiksi.

Vasta saatuani Lapissa käydessäni käsiini muutamia palasia erisävyistä karhunkarvaa aloin olla tulokseen tyytyväinen.

Tämä karva on voimakkaasti kähertyvää ja osittain läpikuultavaa, joten se taittaa auringonvalon eri sävyiksi lisäten elävyyden ja liikkeen vaikutelmaa.

Samanaikaisesti kun kehittelin tätä tummaa kuivaperhoa, olin kokeillut saamieni mullojen syönnösnäytteiden pohjalta eräitä uusia vesiperhosimitatioita märkäperho-ongintaan. Eräällä vaaleahkolla sidonnalla, jonka sävyt ovat suunnilleen kuin vaalean March Brown'in, olinkin jo saanut tuloksia. Halusin lisäksi löytää käyttökelpoisen jäljitelmän värikkäämmistä isokokoisista vesiperhosistamme, lähinnä perhosesta nimeltään Potamophylax latipennis, "large cinnamon sedge". Tämä perho tunnettiin aikaisemmin toisella nimellä, jonka Juhani Aho eräässä lastussaan on tehnyt meilläkin tunnetuksi: Stenophylax stellatus . . ., ruskeasiipi, keltaruumis, iso ja lihava".

Aho käytti tämän perhosen esiintyessä Stone fly'ta, March Brown'ia, August dun'ia tai Orange mallard'ia.

Nykyisin annetaan kirjallisuudessa hyönteisen värisävyiksi oranssinruskea keho ja jalat sekä täplikkäät keltaruskeat siivet. Tältä pohjalta kehitin oman jäljitelmäni: runko oranssin, keltaisen ja vaaleanruskean sekoitusta, siivet lämminsävyistä täplikästä kalkkunaa, runko- ja kaulahäkilät punaruskeaa kukkoa.

Tämä "Orange sedge'ksi" nimittämäni sidos osoittautui heti erittäin hyväksi, ja niinpä siirsin sen värisävyt karhunkarvasiipiseen kuivaperhooni. Tulos oli edelleen myönteinen, jopa siinä määrin, että jouduin saalis-

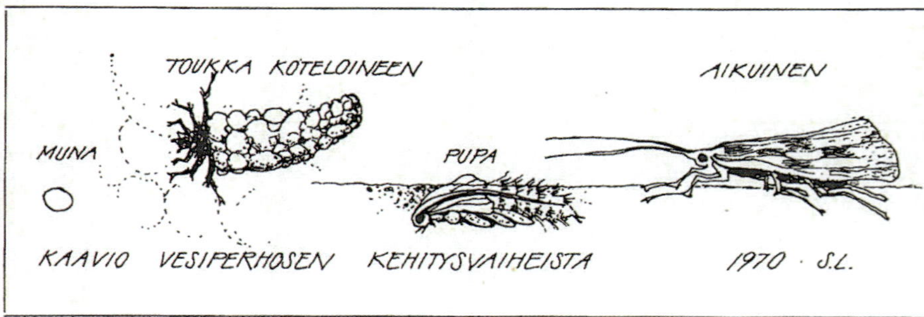

18 URHEILUKALASTUS

The 1970 article that Simo wrote (© Simo Lumme)

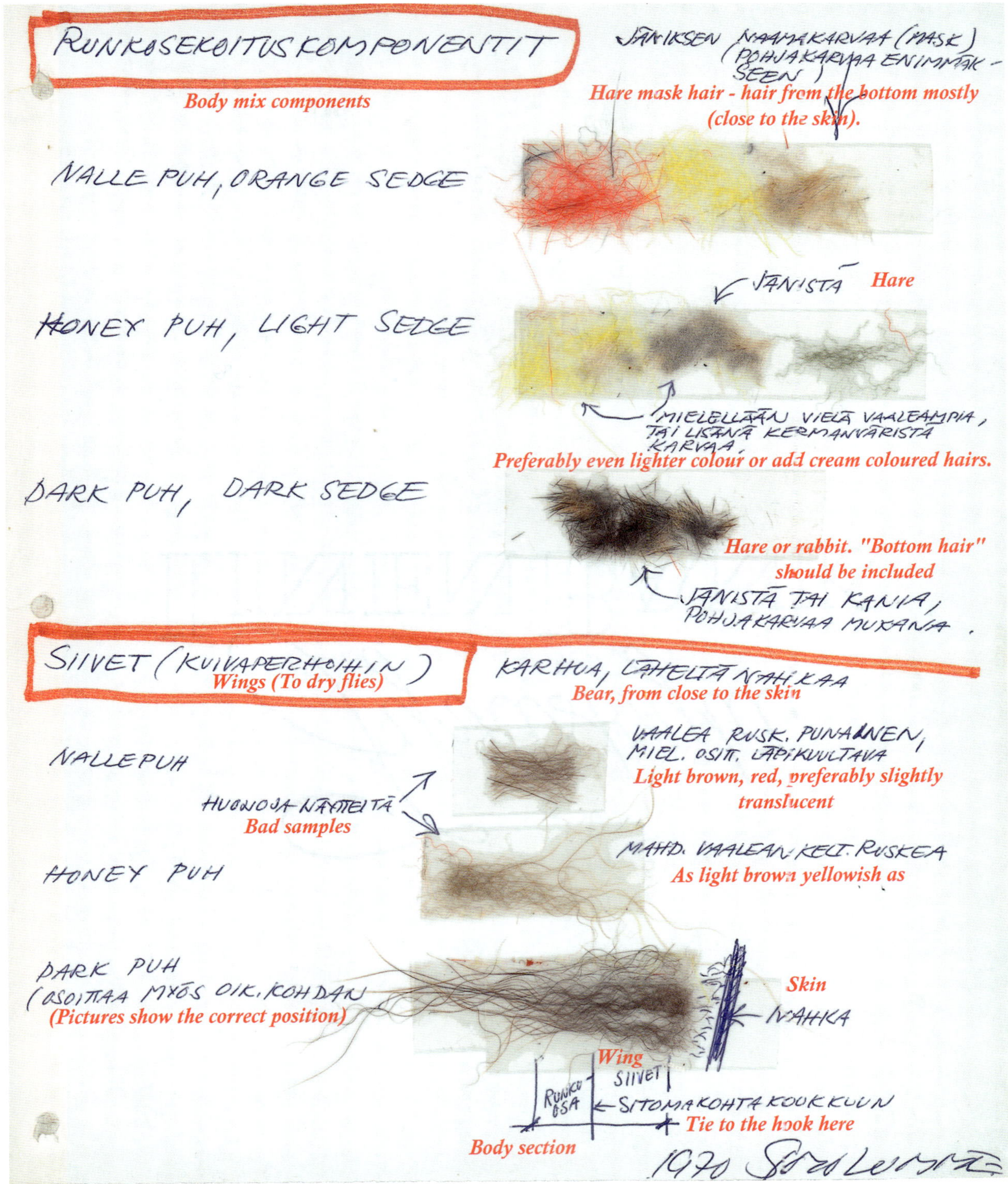

RUNKOSEKOITUSKOMPONENTIT
Body mix components

JÄNIKSEN NAAMAKARVAA (MASK) (POHJAKARVAA ENIMMÄKSEEN)
Hare mask hair - hair from the bottom mostly (close to the skin).

NALLE PUH, ORANGE SEDGE

JÄNISTÄ Hare

HONEY PUH, LIGHT SEDGE

MIELELLÄÄN VIELÄ VAALEAMPIA, TAI LISÄNÄ KERMANVÄRISTÄ KARVAA.
Preferably even lighter colour or add cream coloured hairs.

DARK PUH, DARK SEDGE

Hare or rabbit. "Bottom hair" should be included
JÄNISTÄ TAI KANIA, POHJAKARVAA MUKANA.

SIIVET (KUIVAPERHOIHIN)
Wings (To dry flies)

KARHUA, LÄHELTÄ NAHKAA
Bear, from close to the skin

NALLEPUH

VAALEA RUSK. PUNAINEN, MIEL. OSITT. LÄPIKUULTAVA
Light brown, red, preferably slightly translucent

HUONOJA NÄYTTEITÄ
Bad samples

HONEY PUH

MAHD. VAALEAN KELT. RUSKEA
As light brown yellowish as

DARK PUH (OSOITTAA MYÖS OIK. KOHDAN)
(Pictures show the correct position)

Skin
NAHKA

Wing
SIIVET

RUNKO OSA
Body section

SITOMAKOHTA KOUKKUUN
Tie to the hook here

1970 Simo Lumme

A page of Simo Lumme's most amazing letter that he sent to Lauri Sälevä in 1970 (My translations in red!)
(© Simo Lumme)

wing variation: Klinken's Nalle Puh and it was fly number 25 from my large LT series that I developed during the 1980s in Scandinavia. Most of them were tied in our cosy little cabin on the banks of the Glomma River that we have rented from a farmer for many years.

Regarding my own version of the Nalle Puh I had two variations, one with and one without a tail. The tail version was purely experimental but sometimes worked well during high water conditions in the Glomma in the first three weeks of June. Both were top flies among my wild water

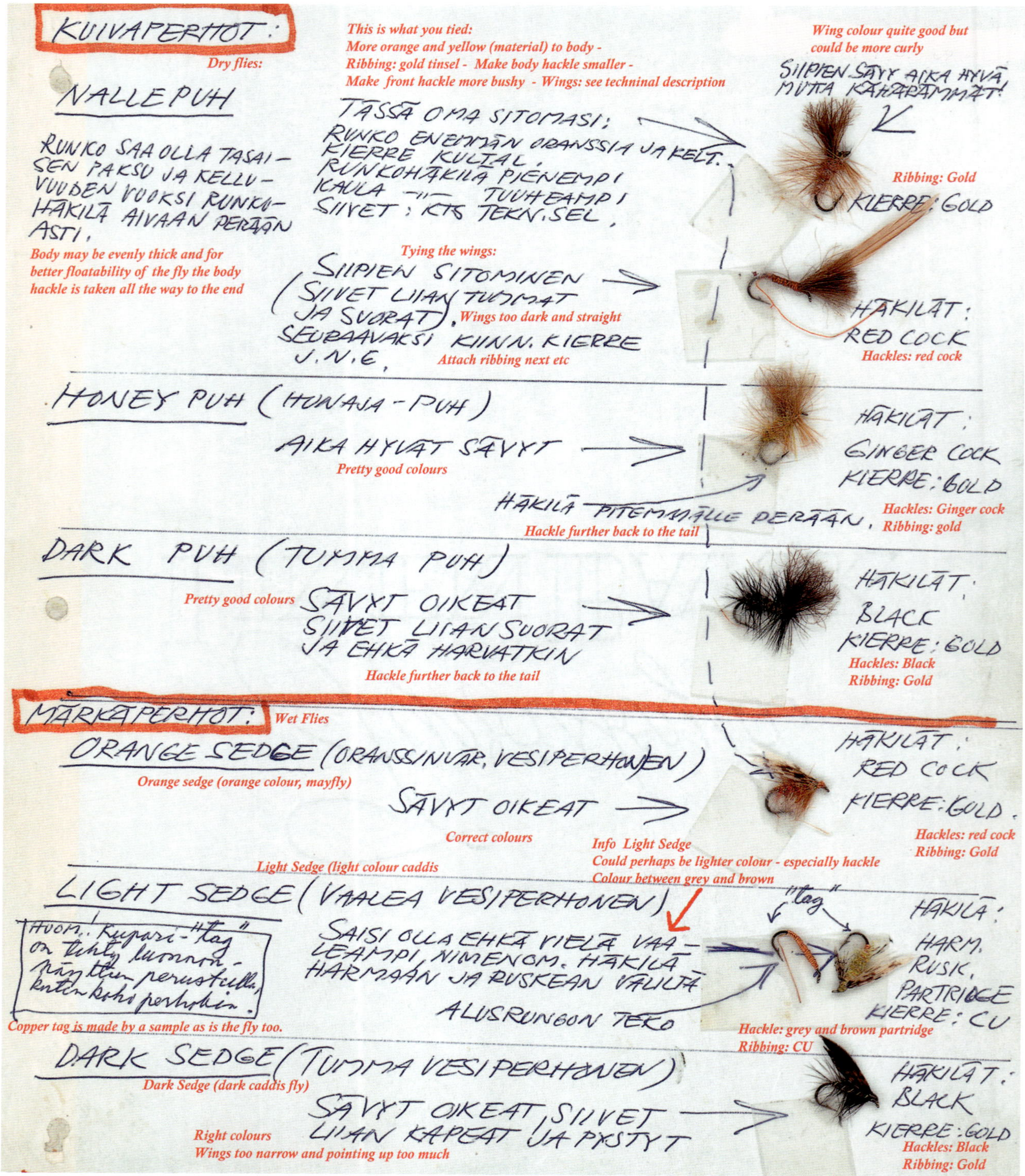

It seems Lauri sent his flies to Simo for a serious evaluation (My translations in red!) (© Simo Lumme)

patterns that were effective before I developed my first Klinkhåmer Special. I often changed the light tan poly dubbing colour depending on which country or area I was fishing.

I got my first Nalle Puhs when I fished in northern Finland sometime in the mid 1970s. I got them from a very nice guy by the name of Pekka Mikkola, whose nickname was Pekkola which is why I remember his full name. I believe he came from the town of Turku in the southwest and spent his summers in the north. He was only a few years older than me and as crazy as I am about nature, wildlife and extremely large dry flies. We met in Rovaniemi and he invited me for dinner and we had such a good chat that after dinner he asked me if I would be interested to spend a few days fishing with him. Never invite a Dutchman because we hardly ever say no! We had a great time together and I was still a rookie with only four or five years of flyfishing experience. I learned a lot while we stayed in the cabin from his parents and we fished three different rivers. When we parted he gave me a box full of Nalle Puhs, Honey Puhs and Dark Puhs and they lasted quite some time. From Pekka I knew that the original fly had wings that were made from brown bear hair and the flies he gave me were indeed all tied with real brown bear hair.

A friend tied them for him and he had boxes full of them. He was a great angler and caster but didn't tie his own flies. He even taught me a few casting tricks when my long distance casting was a real disaster, probably because I didn't need to cast very far to catch fish. From him I also learned that casting and catching fish were two totally different things. On our last day when I caught a few more fish than he did, he told me I was a great flyfisher but a poor caster, especially for distance and I still use his expression today when people tell me I have a funny cast. I remember well that he envied me for my long downstream drag-free drifts and said he had no clue what kind of casts I made, but they were very effective.

When I finally made my own Nalle Puhs, several years later, I used cream and light brown poly yarn as wing material. For the body I used extra fine poly dubbing because I wasn't able to get brown bear hair and original baby seal's fur. So I actually changed the original pattern because I wasn't able to get the wing and body materials which were quite new to me in those days. Luckily, poly yarn turned out to be a perfect substitute for the wings and it prevents commercial and illegal selling of patches of real bear hair too. The high floatability of poly yarn and poly dubbing in combination with a good floatant also helped to keep the Nalle Puh flies floating even longer.

The first tying demonstration of the Nalle Puh

Another Finnish guy who I met in Norway in 1988 at the Fosen peninsula was the first person who showed me how to tie the Nalle Puh properly. His

Honey Puh (left) and Nalle Puh tied by Lauri Sälevä

Dark Puh tied by Antti Vanninen

name was Niilo, but I don't know his last name, simply because he never mentioned it to me. We stayed at the same camping place where we had rented cabins almost next to each other. Because we both fished dry fly only for Atlantic salmon we quickly became friends and one evening he invited us for typical Finnish sausage barbecue, his own homemade beer and Koskenkorva, a very traditional Finnish vodka. He was alone and claimed that setting up a barbecue for just himself wasn't cost effective or cosy. It was turning low tide and we were only fishing a large but superb sea pool and we fished it mainly when the tide came up or was high. As soon the tide went down, all the fish went through and we had to wait for the next run. It was a beautiful spot that I fished for 10 years in a row and knowing the pool and every rock was crucial because fish could only pass through this natural bottleneck when the water was at its highest level.

At high tide the surface was very smooth and perfect for dry flies. He used Bombers and Nalle Puhs and I was very successful with my Klinkhåmer and variations. During our low tide breaks we took a nap, just relaxed or tied flies. After the barbecue we compared our fly boxes and quickly set up our tying stuff. We had probably a few drinks too many, so we were wise and didn't go out fishing that evening. Instead we just compared flies.

When I showed him a few of my own tied 'synthetic' Nalle Puhs and showed him how well they floated, he was quite amazed at how they floated so well without the original materials. He told me that he had lots of real bear hair and seal's fur to tie original Nalle Puhs in his flytying box, and so he became the second Finn who gave me a box with originally tied Nalle Puhs. He also gave me a piece of brown bear and some seal to make it possible for me to tie the Nalle Puh, as it should be tied. In return I gave him a private tying demonstration on how to make a Klinkhåmer and left him a handful of bigger Klinkhåmers that were suitable for catching Atlantic salmon. I also left him some hooks, poly yarn, poly dubbing and most importantly, some of my new floatant, of which I had plenty. It was a great swap for both of us. Unfortunately I could only tie around thirty flies from the bear patch, so I used it very rarely to be sure I still had original brown bear in my collection of special tying materials back then.

What impressed me most about his demonstration was how he tied the palmered hackle and ribbing in perfect harmony. He used a completely different technique for wrapping ribbing over a palmered hackle than I had seen. When I was studying the tying of the Nalle Puh for this chapter, I found out that this method was precisely to Simo's descriptions that I own today. Simo's tying explanation and drawings were totally unknown to me in 1988, so I had no knowledge of that technique, but I knew straight away that I would be using it for ribbing some of my large 'Wildwater' sedges that, beside my Klinkhåmer variations, I used a lot in Norway. I also had lots of confidence that it would greatly improve the durability of all my bushy-tied palmered flies in which I also use ribbing. I even added ribbing to flies that had none, and that's how Simo's ribbing technique pretty soon made many of my palmered hackle flies a lot better and longer-lasting in the late 1980s and early 1990s. The third set of Nalle Puh flies I got from my close friend Antti Vanninen.

Nalle Puh materials described in more detail

Today there is a whole family of Puh flies: Nalle Puh, Dark Puh, Honey Puh, Snösko (Snowshoe) Puh, Aippa Puh and CDC Puh. I am sure that there are many more variations worldwide that are nameless or have totally different names. There are a few basic rules in tying the Puh series. For example: Always keep in mind that when you are tying the Nalle Puh series that the wings should be more upright than forward, and the body hackle fibres should be slightly shorter than the ones in front of the hackle. Forward wings should be avoided as they topple the fly onto its side. I read this a few times in Simo's notes and it became very important to me. These two vital tips also help to reduce the chances that the fly will float face or front down in the water.

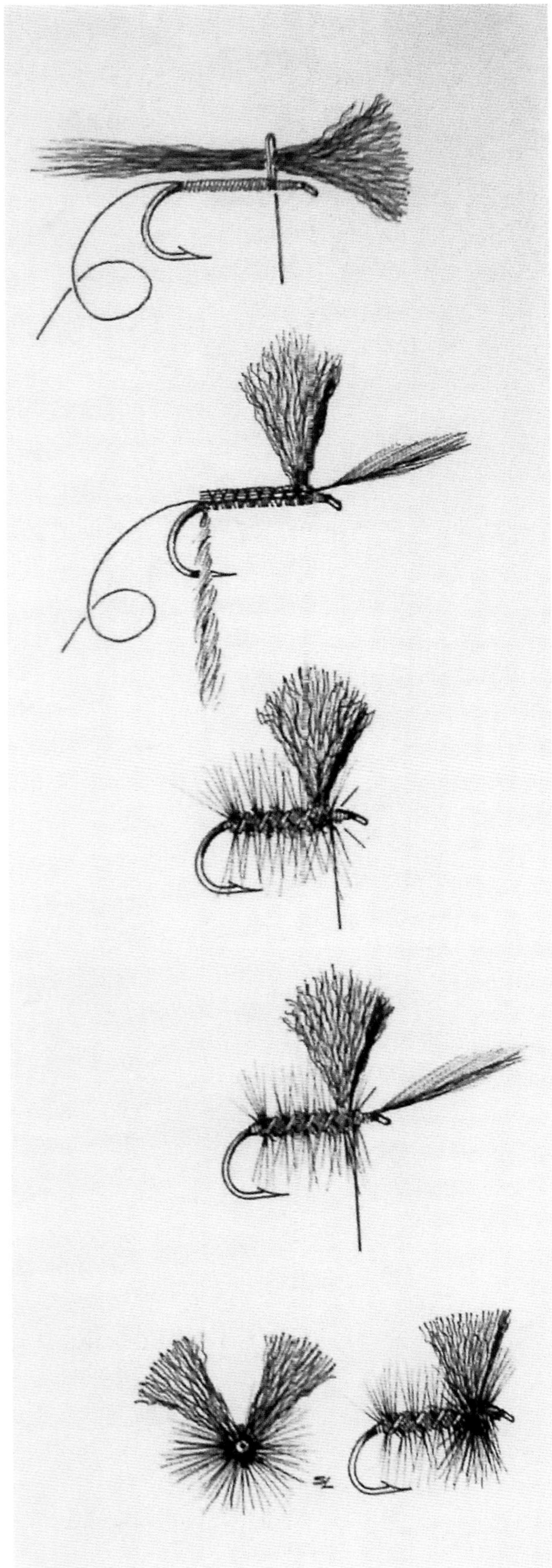

Simo's original drawings from his book Perhonsidonta
(© Simo Lumme and Juha Pusa))

Simo used Permaflote heavily like so many flyfishers in the 1970s and 1980s. Without doubt it was the best liquid fly floatant ever invented. With Permaflote, wing and tail fibres or hairs never stuck together and it also worked outstandingly well with poly yarn and poly dubbing because of its absorbing qualities.

Then suddenly it was no longer available, probably due to the fact that it contained carbon tetrachloride, a very dangerous chemical, considered to be highly carcinogenic. After it was banned, several fake substitutes came on the market in liquid, gel and even spray form. One of the replacements was named 'Permafloat', but it was less effective compared to the original. The original Permaflote was developed by Richard Walker and chemist Arnold Neave and was distributed in the UK by Chubbs Ltd from Amersham and was only available in a little glass bottle. I personally used it until the late 1980s because it was almost magically effective and kept flies floating like cork. Simo submerged his flies for 24 hours or even more in Permaflote and because that process darkens the body a little, he used slightly lighter body colours. My only observation of a downside with this floatant was that the chemicals within Permaflote dissolved the varnish on my flies that was, and still is, a mix of cellulose diluted with acetone. That's why in the old days I hardly used varnish on my dry flies, just tied in a few extra knots. This was important to me because when too much varnish had been dissolved in the chemicals and you then dropped in your flies, when the Permaflote evaporated, some materials could be left hardened or lose their typical softness. Sometimes colours could also be changed by Permaflote, which explained to me why misunderstandings can occur when people want to copy a fly and misinterpret the correct colour combinations.

Although the Puh series of flies look simple to tie, it is still easy to make mistakes, especially in the dimensions or quantities of material. Incorrect proportions happen easily, resulting in bad floating and wrong landing flies while casting. A well-tied fly lands perfectly on the surface with the wings upright and continues its drift in the best possible way. I therefore will split up the tying instructions into separate parts.

Honey Puhs tied by Ariko Miettinen

Antti Vaninnen with a very nice brown trout caught with Nalle Puh (© Antti Vanninen)

Hooks for Nalle Puh

For the hook, Simo recommended size 6-12 dry fly down-eye hooks. Most common was size 8, because that size fits better to the fly sizes that Simo tried to imitate. In the 1970s Simo mainly used Mustad hooks because Partridge hooks were hardly available in Finland. On the cover photo from Simo's and Juha's excellent flytying book you will see a Nalle Puh that is tied on a dry fly Mustad down-eye from the late 1960s.

There are several dry fly hooks with a nice wide gape suitable to tie a perfect Nalle Puh, but most tyers use their own preferred hooks. When Partridge hooks became available Simo also used Captain Hamilton dry fly and down-eye hook L3A in the largest sizes. In Simo's letters, I also saw that he mentioned a Partridge Wide Gape Trout hook.

I love big flies, and believe me, you can make the Nalle Puh really big, especially when you use the pattern in remote and subarctic countries. My long-time experience is that a lot of people tie their flies far too small when they fish in Scandinavia, Russia, Alaska, Canada, Mongolia and even in tropical areas. Of course, you will catch a lot of fish using size 14 and 16 Nalle Puhs, but those who do so will not know how well they might have done using sizes 8 and 6. In general, you catch fewer fish when using bigger flies, but larger fish; at least that's my experience. However, when fishing very remote places, the catches on flies tied on hook size 8 and 6 can reach crazy numbers too. I have tied Nalle Puhs up to hook size 1 for Atlantic salmon and for Lenok trout in Mongolia and it worked very well. In Scandinavia in early June I never used smaller Puhs than size 6 and in July I went down to size 8 which is the smallest size I tie my Nalle Puhs, and that's how I caught my whoppers. In August and September when water temperatures are much higher and

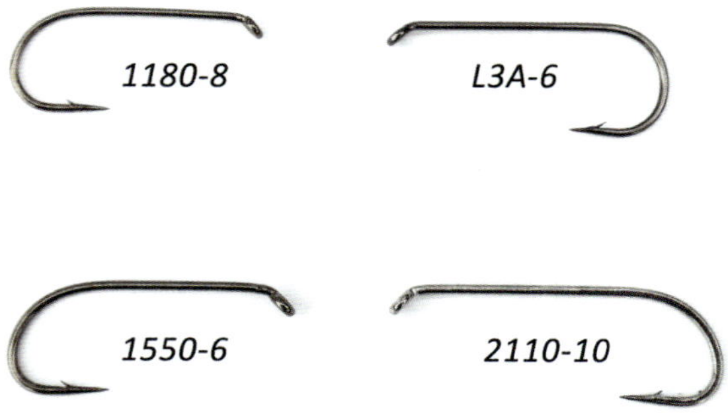

the rivers are lower I use even smaller flies in Scandinavia, but those are completely different patterns and more realistic too.

For my synthetic wing version of the Nalle Puh series I personally use the Daiichi 1180 size 8 and for my original version or big synthetic versions I will use the Daiichi 1550 size 6 or Daiichi 2110 size 10 which are exactly the same hooks that I use to catch my big salmon on dry flies. I also use the 1550 hooks when I fish rivers in the Yukon, Alaska and Mongolia where the fish are usually bigger, and larger dry flies are the key to success. The Daiichi 1550 size 6 is just slightly bigger than the 1180 size 8, but has a stronger wire, which is why I use that particular hook for the bigger fish. A welcome benefit of the 1550 hook is that it is a bit heavier so it is less likely that flies will float on their side when they land on the surface. After using the Puhs for many years I also discovered that Nalle Puhs tied with a longer body can be deadly in a lot of rivers all over the world. The Daiichi 2110 is actually a Bomber and Bug hook with a longer shank and fits perfectly for that particular situation.

I have no quarrel with people who use smaller hooks than I do and I have a few friends who only use Puhs tied on size 10-16, but maybe they go for numbers of fish instead of trying to hook a real whopper as I do. One of my friends only ties synthetic versions in size 16 and 18 and is very happy with these patterns even in fast-flowing rivers. Sadly, he has never fished in Finland, nor in any other countries in Scandinavia, so for him a size 12 is already a monster fly.

Tying silk

Simo used tying silk instead of normal tying thread and for the Nalle Puh he recommended hot orange. I am quite sure he used number 19 hot orange Pearsall's Gossamer. Preben Thorp Jacobsen mentioned it in his story about the Nalle Puh as well, after he had contacted Simo. For the Dark Puh it is simple: it would have been black Pearsall's Gossamer no 9; and for Honey Puh no 16 or 20, in olive or light olive. I love silk as tying thread and use it a lot for my large dry flies and all my soft hackle flies. I also use pure silk threads a lot as a ribbing in most of my dry flies. Sadly the new generation of tying threads are so fine now that a lot of tyers have problems using a 'thick' silk vintage thread from the past.

Wing materials

The bear hair should be cut from close to the skin. Simo liked light brown, reddish preferably slightly translucent colours for the Nalle Puh and light brown yellowish for the Honey Puh and, of course, the darkest hair for the Dark Puh. Sadly, good quality brown bear hair is almost impossible to get. Having bear skin without a legal EU certificate is an offence in Europe.

To get my legal patch with official certification took many years. This is also why there are so many Puhs tied with synthetic wings. In my opinion synthetic poly yarn and polypropylene

Some of the silk threads that I use when tying Nalle Puh flies

fibres are the best substitute materials to replace the natural bear hair as wings. I have been using poly yarn and polypropylene for nearly forty years and can say with great confidence that I am very satisfied with it as wing and wing post material. Frizzy poly yarn replaces the curly bear hairs in an excellent way and for Simo the curly ability of the bear hair was very important in his thoughts behind his pattern. Beside that, poly yarn is very cheap, widely available and almost unsinkable when treated well with a good floatant. In catching results between original and synthetic hair and body materials, I hardly found any differences. Sometimes a fly with the original material was more successful while in other rivers I did better with a synthetic version.

As I have said many times before: I would never kill an animal to use its fur and feathers for my flytying, but I don't mind using natural materials privately when it's a road kill or the animal passed away by a natural death.

The exact location of the wing and where to tie in the hairs

The wing is essential for volume, length, and in particular in regard to its attachment point. A too long wing, or attaching the wing too close to the eye of the hook means your fly can easily land face down in the water. Attached too far back and the entire fly can go out of balance during casting, landing or floating. To find the right proportions is not easy in the beginning so intensive testing of the completed flies is a very good idea to get the amounts correct and find your way.

The wing colours and how to tie in the hair properly were extremely important for Simo. In his written wing samples that he sent to Lauri in 1970 he even included some bad examples. The intention of his letter and including real material in it, explained the correct positions in the best possible way. With an extra drawing he even showed Lauri where to cut the bear hair from the skin, what part to use as the wing, where to tie the hair in on the hook shank and what part to use as underbody. I think it is the best and most important example Simo ever made.

Body materials and mixtures for the Nalle Puhs

When I discussed body colours more fully with Juha he confirmed that there were some differences in body mixtures and their colours as well and he believed that they have changed slightly from early years, probably because of

(© Simo Lumme)

Permaflote treatment. I also read and heard about some different dubbing materials and mixtures that tyers were using for Nalle Puh bodies which resulted in the dubbing becoming an important part of the Nalle Puh for me. Another reason is that since the early 1980s I have made a lot of dubbing material myself and it has become an important hobby within my flytying for the last 40 years. I actually started to first make my own dubbing by cutting my flytying waste into small pieces in the 1980s and putting them into a coffee grinder. It worked well especially at a time when I didn't have a lot of money and many dubbing and other flytying materials were hard to get. To be honest, I actually split the body materials of the Nalle Puhs into three parts, namely:

1. the original body materials as used by Simo and his friends
2. existing synthetic dubbing materials as substitute for seal's fur
3. high floating experimental dubbing mixtures as I use in most of my rough water patterns

Let's start with the original body materials of the Nalle Puh, Honey Puh and Dark Puh.

In the letter where Simo showed his perfect mixture to Lauri he used following materials, colour combinations and amounts:

1. Nalle Puh or Orange sedge: 1/3 orange seal's fur, 1/3 yellow seal's fur, 1/3 hare (exactly as in the book). The hare dubbing that Simo used was taken from a hare's mask cut from the bottom close to the skin!
2. Honey Puh or Light Sedge: 40% yellow seal's fur, 40% light coloured hare, 20% green seal's fur.
3. Dark Puh or Dark Sedge: the dubbing for the Dark Puh was mainly hare or rabbit from which the bottom hairs were included.

An important note that Simo mentioned to Lauri about the body shape was that it must not be too long or cone-shaped, but even width at all times. This is the main mistake I made myself in the past and I also see it a lot nowadays in copies of the Nalle Puh globally.

Regarding the Dark Puh, I am personally not a big fan of using rabbit hair in dubbing for dry flies because it absorbs water too easily. However, when using Permaflote in the past as Simo did, it made rabbit quite floatable. Several of my Finnish friends use a mixture of black and brown seal's fur or dark brown seal's fur mixed with hare.

When it comes to synthetic substitutes for Simo's seal's fur and hare mixtures, Fly-Rite extra fine dubbing suited me well for many years. For the Nalle Puh for example, I liked to mix Golden Brown (#17) with a tiny bit of Rusty

Orange (#18). For the Dark Puh I used Dark Grey (#7) or Black (#2) and also mixed those two. Finally, for the Honey Puh I used Tiny Blue Winged Olive (#24) or Blue Winged Olive (#10) or mixed both as well.

Today my preferred dry fly dubbing is a mixture of a highly floatable synthetic polypropylene mixed with 60 year old seal's fur and hare. With my special coffee grinder I can make it as fine as I want it. If I can't make the desired colour I grab another wonderful dubbing material. It's a natural material known as Kapok. Since 1996 I have used Kapok (*Ceiba pentandra*) in flies that I really want to be unsinkable.

The body mixes as suggested by Simo in his letter to Lauri.
From top to bottom: Nalle Puh, Honey Puh and Dark Puh (© Simo Lumme)

My personal set of Nalle Puhs made from a mixture of 60-year-old seal's fur and hare from mask

Ribbing

In many publications about the Nalle Puh you will read about all kinds of ribbing material and colours, but Simo only used gold wire as the rib for all his versions.

Front and body hackle

In the late 1960s and early 1970s the hackles were lot different to the quality we enjoy today. The hackles were often shorter and the fibres much longer and softer. Simo also tied body hackle all the way down to the end of the body. Extremely important for Simo was that the body hackle fibres were a bit shorter than those from the front hackle. He used the following hackles: for the Nalle Puh a red cock hackle; for Honey Puh a ginger cock hackle and for the Dark Puh a black cock hackle.

In the past, you sometimes needed several hackles for one Nalle Puh while nowadays you can easily tie a Puh very bushy with only one long hackle. That's probably why I have seen several people using only one hackle instead two or three, as was done in the early days. I know Simo cut hackle fibres at the bottom if they got too long. He showed that clearly in his letter with the tying step drawings he sent to Lauri where he marked a cutting line equal in width to the gape of the hook.

Simo also advised cutting the front hackle directly between the wings to better emphasise them. You can see this clearly in the drawings in his book where no hackle fibres show between the wings.

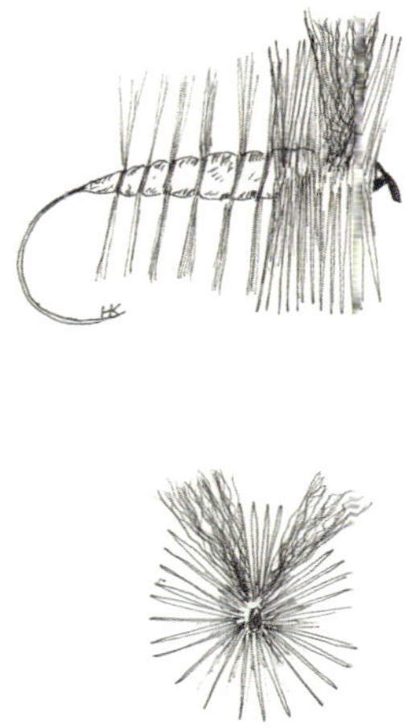

No 25 Klinken's Nalle Puh

My White Puh as I tied them in the early 1990s

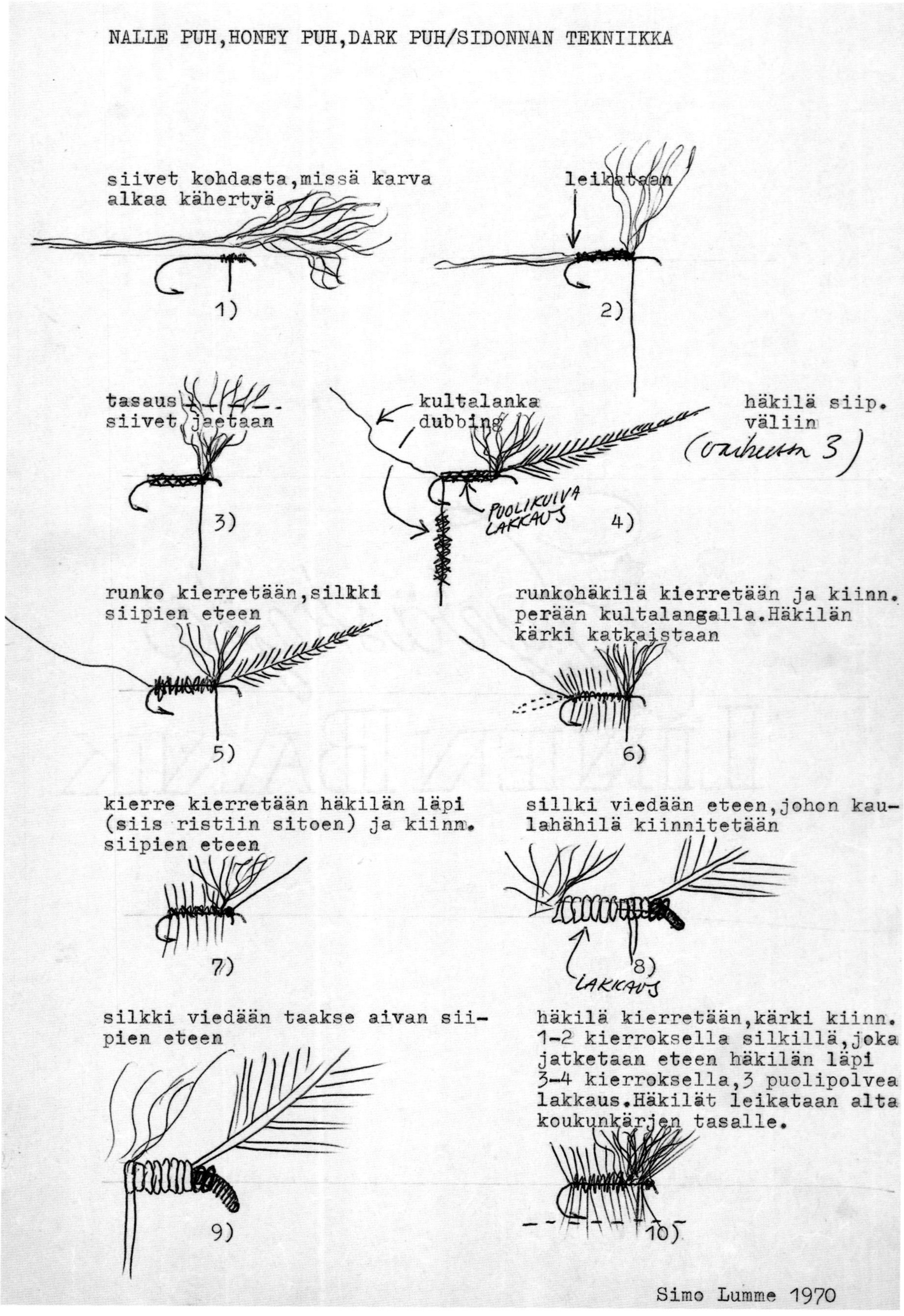

Nalle Puh, Honey Puh, Dark Puh general tying instructions as suggested by Simo Lumme to Lauri in 1970

Tying the original Nalle Puh

Original Nalle Puh (1968-1969) tied in the spirit of Simo Lumme

Dressing

Hook:	Mustad down-eye size 6-12 (most common was size 8)
Thread:	Pearsall's Gossamer No. 19 - hot orange
Wing:	Light brown bear, red preferably, and slightly translucent
Ribbing:	Fine gold wire (counter wrapped over hackles on the abdomen)
Body:	A mixture of orange and yellow seal's fur mixed with hare's mask blend
Body hackle:	Good quality red-brown, wrapped from front to back secured with ribbing
Front hackle:	Reddish brown rooster with longer fibres than body hackle
Head:	Pearsall's Gossamer No. 19 – hot orange

Tying steps

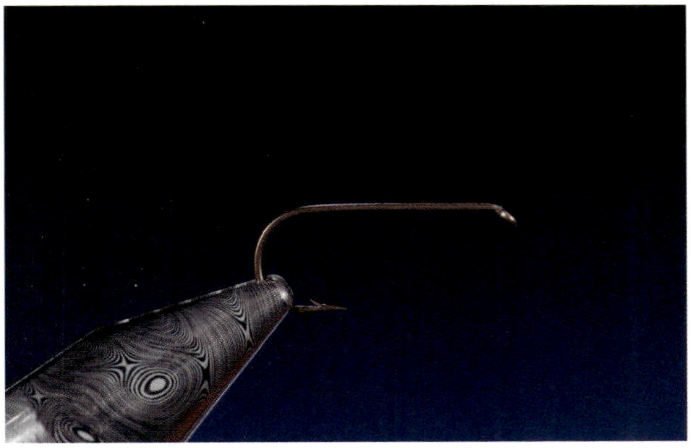

1 Pinch the barb off the hook if you wish and place the hook in the vice.

2 Set up your thread.

3 Tie in the wing about ⅓ from the hook eye. Fasten the hair where the curly part begins. *The part cut from the skin is in front!*

4 Wrap the silk up and down to the bend to secure the hair and cut off surplus. Bring the silk back to front and tie up the wing.

5 Split the wings in two equal parts. I am using figures of eight and I also wrap the thread around the stem of both the wings.

6 Diagonal view.

7 Front view.

8 Cut the wings to required length and tie in the ribbing underneath the hook shank. Then tie in the body hackle in front of the wing.

9 Bring the silk thread to the end of the body and apply the dubbing material on the silk. Dub the body forwards to the wing.

10 Wrap the hackle backwards to the end of the body and secure it with the golden wire.

11 Counter-wrap the ribbing through the palmered hackle and secure it in front of the wing. Trim the excess off the hackle.

12 Tie in the front hackle.

13 Wrap the front hackle and wrap the silk through the hackle with three or four turns and whip finish.

14 Trim the excess off the hackle and varnish the head. Simo trimmed the hackle under the hook on the level of the hook point. (*See his drawing from 1970.*)

15 Front view. Nowadays many tyers use only one hackle because they have developed so much since the original was made.

16 Simo also cut away the hackles between the wings and the Nalle Puh is completed.

More dressings

Dark Puh (original version by Simo)

Hook: Mustad down-eye size 6-12
Thread: Black Pearsall's silk No 9
Wing: Bear hair – as dark as possible
Ribbing: Gold wire
Body: Hare with bottom hair included
Body hackle: Black rooster wrapped from front to back
Front hackle: Black rooster with longer fibres than body hackle

Honey Puh (original version by Simo)

Hook: Mustad down-eye size 6-12
Thread: Light olive or olive Pearsall's silk No 16
Wing: Bear hair – as yellowish light brown as possible
Ribbing: Gold wire
Body: Yellow seal's fur, green seal's fur and light hare
Body hackle: Good quality ginger rooster wrapped from front to back
Front hackle: Ginger rooster with longer fibres than body hackle

CDC Puh (originator: Juha Pusa)

Hook: No 6 Partridge Capt Hamilton Dry Fly
Thread: Orange
Wing: Greenish-brownish or tan/light/medium brown, a pair tied in a V-shape
Ribbing: Clear monofilament, 0.15mm
Body: Orange poly dubbing with a thin layer of olive poly dubbing over
Body hackle: Red/brown rooster
Front hackle: Tightly tied red/brown rooster

The idea of using CDC is because this pattern is very large and also quite heavy with bear wing. CDC makes it much lighter and it floats and skates better.

Snösko (Snowshoe) Puh (originator: Juha Pusa)

Hook: No 12-14 Partridge Hamilton Dry Fly
Thread: Orange
Wing: Light hair from the foot of a snowshoe hare, tied in a V-shape
Ribbing: Clear monofilament, 0.15mm
Body: Orange and olive poly dubbing mixed, tightly dubbed
Body hackle: Dark red/brown rooster
Front hackle: Same as body hackle wrapped tightly

My Nalle Puh and Honey Puh tied with original materials

My own version of the Nalle Puh

Hook: Daiichi 1180 size 8 or Daiichi 1550 size 6

Thread: Pearsall's Gossamer No. 19 – hot orange

Wing: Brown bear hair tied in upside down in an upward V-shape

Ribbing: Fine gold wire (counter-wrapped over hackle on the abdomen)

Body: A mixture of orange and yellow seal mixed with hare's mask's blend (nice and buggy)

Body hackle: Good quality red-brown, wrapped from front to back

Front hackle: Reddish brown with longer fibres than body hackle.

Head: Pearsall's Gossamer No. 19 – hot orange

My Dark Puh, tied with original material (left) and with synthetic wing (right)

Tying my version of the Nalle Puh with synthetic wings

Dressing

Hook: Daiichi 1550 size 6 and for special occasions the Daiichi 2110 size 10
Thread: Semperfli Pure silk, Vintage Orange (6a)
Wing: Cream poly yarn, tied in an upward V-shape
Ribbing: Fine gold wire
Body: A mixture of orange and yellow polypropylene and hare's ear mixed in coffee grinder
Body hackle: Good quality red/brown, wrapped from front to back
Front hackle: Reddish brown with longer fibres than body hackle.
Head: Same as tying thread

Tying steps

1 Pinch the barb if you wish and put the hook in the vice and set up your thread.

2 Cut two pieces of poly yarn, make loops and taper the ends with your scissors to about 45 degrees. Tie them both on the hook shank about ⅓ from the hook eye.

3 The loops make it a lot easier to split the wing in two equal parts. Use your thread to separate the wings and tie the wing up in vertical position.

4 Front view.

5 Split the wings well and secure each of them firmly with your silk thread. Then cut off the loops and cut them to required length.

6 Tie in the gold wire and bring your thread to the back.

7 Attach dubbing to thread, make a dubbing loop, twist the dubbing and wrap your body.

8 Tie in your body hackle between the wing parts.

9 Top view.

10 Wrap the hackle backwards to the end of the body and fasten it with the rib. Secure body hackle with the ribbing wire with counter wraps through the palmered hackle forward, secure and snip off.

11 Tie in the front hackle.

12 Wrap a front hackle and create a nice collar front hackle.

13 Front view completed fly

Fishing the Nalle Puhs

In his article Simo explained clearly that his Puh series isn't very good in slow-running rivers. It's a dry fly that is especially designed to be fished in fast broken water where the current is flowing wildly and forming whirls, which give a life-like impression to the fly. I use the same leader as I describe in my chapter about Poly Bugs.

My set of homemade dubbing for making Nalle Puhs

Chapter 12

Perfect Poly Bugs for Atlantic Salmon

After 53 years flyfishing for Atlantic salmon I must admit that I haven't written much about these wonderful, leaping silver fish, or about the special flies and techniques that I use to catch them. I have written a lot about trout and grayling, and even more about the flies and fishing techniques I use to catch them. My special grayling flies are now so popular that it was always far easier for me to write about them. I have fished a lot for grayling, especially in the 1980s, 1990s, and in the early years of the new millennium. Back in those days, flyfishing journalists often described me as a crazy grayling junky. But the reality is that I put in many more hours fishing for Atlantic salmon than for any other species.

Perhaps the main reason why I was so shy to share my knowledge of Atlantic salmon was because of a bad experience in my early twenties. At that time I was still a greenhorn who enthusiastically wanted to share his stories and experiences about catching salmon and seatrout on dry flies in the most northern parts of Norway, and nobody believed me.

Previously, dry fly fishing for salmon was not popular in Europe, and a lot of people even called me a liar and claimed that it was impossible to catch a salmon or seatrout on a dry fly. Those words were not just uttered by unknown people, because a few among them were quite famous, and at the time were roughly the same age I am today. Fortunately, this unhappy experience had an enormous positive impact on my flyfishing career, because it forced me to isolate myself and continue to do what I was doing and what I believed in. So I went my own way which, luckily, has always worked for me. It also taught me never to make the same mistake with young people who wanted to share their experiences and knowledge with me, and maybe that's why I have done so well with the many kids' classes and workshops I have organised over the last nearly 25 years.

A crowd of Norwegians witness my dry fly salmon fishing

Being a dry fly purist for Atlantic salmon was tough in Europe in the 1970s and even early 1980s, and made me feel a complete loner. But today I can proudly say that it was the perfect path for me to follow. The second problem I had was the fact that the rivers I was fishing were all located in the most northern part of Norway and Finland and by train that was a long trip. So when my student days were over I changed my fishing haunts to focus more on fishing in south and central of Norway.

In 1988, after sixteen years using mainly dry flies for Atlantic salmon, I finally got recognition when I proved, in front of nearly a hundred people, how I hooked and landed fifteen salmon between 5pm and 11pm with only one particular dry fly, the Klinkhåmer Special. This all happened at the Fosen Peninsula just below the Støvelfossen in the Stordalselva in Åfjord on 7 July, when some people watched me catching fish after fish, and the news spread faster than wildfire. They couldn't believe what I was doing, and when I

TORSDAG 14. JULI 1988

Tips Fosna-Folket 28 105 - Tips Fosna-Folket 28

Nederlandsk flue-binder-ekspert

Mange fosninger står i disse dager trofast i våre elver med håp om å dra opp den store laksen. Også turister tar del i denne «sporten». En av de er nederlenderen Hans van Klinken som er spesialist i fluebinding, og som skriver artikler for ulike magasiner i flere land om elvefiske.

F-F: Terje Dybvik

Fosna-Folket treffer Hans van Klinken og kona Ina på Austråt Camping. Hans er i Norge for 20. gang i år og i Fosen for andre gang. I år er han i oppdrag for turistnæringen i Norge, for å lage informasjon for andre turister om fiskeplasser og steder hvor det er å fint å campere. Mye av informasjonen som tidligere har vært gitt til turister har vært fra folk innen turistnæringen som ikke har hatt samme forhold til fiske som Hans van Klinken.

— Fint på Fosen!

Hans er ansatt i det nederlanske forsvaret, men tar tre turer hvert år i utlandet for å dyrke hobbyen sin som fluebinder og fisker. Det dreier seg stort sett om Skandinavia og Storbritania. Samtidig skriver han artikler for ulike magasiner fra flere land, uten å få noe penger for det. Det vil han heller ikke motta, da han gjør dette av ren egeninteresse. Artiklene omhandler spesielle fluer som han har produsert, reiselivshistorier og selvsagt mange gode råd til andre fiskere. Han har gjennom årens løp blitt vel anerkjent for sin viten innen emnet, både i Nederland og i utlandet.

Ina jobber på laboratorium på et sykehus og begge bor nå i en liten by i Nederland som heter Harskamp. De flytter en god del i landet på grunn av at Hans er ansatt i forsvaret. Sammen utgjør disse to et lettlivet par som mer en gjerne tar en prat med Fosna-Folkets utsendte medarbeider. Undertegnede fikk samtidig frisket opp sine språkkunnskaper, som nok hadde ligget i dvale en tid.

Både Hans og Ina er enige om at Fosen er et fint sted. De dro først innom Åfjord, men har likt seg best ved Austråt Camping. Hans sier mange ikke vet om denne campingplassen, da den er utelatt i mange utenlandske turistbrosjyrer. Flere turister tar derfor turen om Åfjord og så rett til Trondheim. Hans skryter av været i Fosen som har vært lett skyet. Det er midt i blinken for en som driver med fiske. Vanligvis fisker han minst to ganger hver dag, men har nå i sommerferien tatt hensyn til at Ina også er med i reisefølget. Hun liker godt å slappe av med en bok, og foretrekker solskinn framfor gråvær. Når det gjelder hennes fiskeinteresse, så fisker hun helst på sjøen. De har med seg en kano til dette bruket. Når det er gråvær padler stort sett Ina og Hans sitter med fiskestanga. Motsatt er det i solskinn hvor det er Hans som padler og Ina ligger bakerst og nyter været. Fin arbeidsfordeling med andre ord!

Hans og Ina van Klinken har vært turister på Fosen noen dager og likt seg godt her. Hans er spesialist i fluebinding, og holder fram en laks han fikk dagen før Fosna-Folket møtte paret. En kveld i Åfjord utklasset han de andre fiskerne ved å dra på land 15 lakser!

Laksevandring på video

Hans har tatt opp et times opptak fra laksevandring i elvene i Åfjord. Det skal han bruke under foredrag i hjemlandet og andre steder. Å stå og se på laksen jumpe oppover strykende er veldig avstressende som flere burde oppleve, sier Hans.

Hans liker best å fiske alene uten altfor mange turister rundt seg. Derfor har han endel hemmelige fiskeplasser som han vil ha for seg selv, også en her på Fosen. Dagen før vi var og pratet med paret, hadde Hans fått en laks på bortimot tre kilo der. Den fikk han på en flue han har laget selv som har fått det norsklydende navnet Klinkhåmer-spesial. Yndlingsfiskesortene er forøvrig laks og harr.

Har han så noen gode råd til fiskefantomene her i området? Ja, en erfaring han har gjort seg er at laksen tar bedre på tørrflue hvis elva er lav.

Uspiselig fisk i Nederland

Hvorfor kan så ikke Hans van Klinken dyrke sin fiskeinteresse fullt ut i Nederland? Jo, der har det nå blitt så mye forurensning at fisken er blitt uspiselig. Miljøverorganisasjonene er veldig restriktive og påbudt at det for eksempel ikke skal være noe mothake på kroken man fisker med. Dermed kan en uten å skade fisken, ta den av og kaste den ut i vannet igjen. Det er så stor motstand mot fiske der nede, at de ikke kan publisere bilder som viser en fisker med sin fangst på en pinne! Også i Norge har det som kjent gått ned med flere fiskebestander, noe også Hans har registrert siden de første gangene han var her. Vi får vel likevel håpe det ikke blir så galt som i Nederland!

— Gjerne tilbake til Fosen

Hans og Ina sier de gjerne tar turen tilbake til Fosen igjen. De har trivdes godt og roser Austråt Camping som har vært midt i blinken for dem. De vil også trekke fram Fosenbrosjyren som de mener er veldig god. Den er informativ og oversatt til flere språk, noe som gjør at turister lett kan gjøre seg kjent i området. Mange andre steder har de savnet nettopp en slik.

Etter vi forlot paret, får vi nå opplyst ved Austråt Camping at Hans 7.juli fra klokka 17-23 tok opp 15 lakser. Den største var på rundt 7 kilo. Andre fiskere rundt måtte bare gi seg over, og godta at fluene til Hans van Klinken var best.

Vi tar farvel med Hans og Ina van Klinken, håper de vil drive god reklame for Fosen og ønsker dem velkommen tilbake!

Denne tørrflua er en av de Hans van Klinken har laget. Den har fått det norsklydende navnet Klinkhåmer-spesial.

Fosna-Folket
Grunnlagt 1964

Redaksjon og ekspedisjon:
71 60 Bjugn
Tlf. 076-28 105
kl. 0800-1600
(mandag-fredag)
Telefax: 076-28 861

Avdelingskontor (annonser og ekspedisjon)
7130 Brekstad
Tlf. 076-24 612
kl. 0900-1400
(mandag-fredag)

Abonnem
Helår
Halvår
Kvartal
Trykk
Fosna-Folket A
7160 Bjugn

released most of the fish as well, it was seen as the action of a lunatic! In those days, catch-and-release was not popular. The largest fish that I landed and safely returned to the river was 87cm (at least 7kg) and I lost an even bigger one that kept me busy for several minutes. Because a local radio station had announced that something really weird was happening below the waterfall, more and more people turned up, and in no time the whole area was packed with people. The low water made the river easily accessible, and I finally had to stop fishing around 11pm because there were simply too many people walking behind me and I got worried about hooking somebody with my backcasts. Not many people had been fishing the previous few days because of hot temperatures and no fish had been caught for some time. Together with a Finnish guy who caught nine fish using a Bomber, we wrote dry fly history that day. The hot weather and low water conditions were perfect for dry fly fishing and we were both very accustomed to fishing in conditions like these. I only kept one nice grilse to put in the freezer and smoke later for friends from Holland.

On 14 July, journalist Terje Dybvik from Fosna-Folket wrote a full-page story in the local newspaper about what happened on 7 July and described it as an extraordinary experience, something that he had never ever seen before. For me it was just one of many similar fishing experiences I had over all those years that I had used dry flies to catch salmon, seatrout and sea-run Arctic char in northern Norway and Finland.

My big change with salmon fishing came in 1988

When I caught my first salmon below the surface in the Bondalselva (still in Norway and also in 1988), I started fishing more seriously for salmon with a wet fly. However, my interest in dry fly fishing for salmon and seatrout so occupied my mind that I began hunting for every article and

Left: Klinkhåmer I use for Atlantic salmon in Canada, Norway and Finland, and right: one for catching grayling

story about Atlantic salmon or seatrout caught on dry flies. My search quickly led me to flyfishers and flytyers who were familiar with flyfishing in Atlantic Canada, New Brunswick, Quebec and even in parts of the State of Maine, USA.

Meanwhile my Klinkhåmer Special had been garnering more international recognition, and as a result I was able to make new contacts a lot easier. Being a participant in the international flytying symposium in New Jersey as one of the European flytyers in the mid 1980s helped me a great deal too. With further help from great friends like the late Alan Bramley from the UK, and the late master flytyer Hans P C de Groot of Holland, I got to know many other well-known flytyers and flyfishers, such as Warren Duncan and Marc Madore from New Brunswick and Al Worthington from New Hope, Pa., USA.

Alan Bramley was the owner of the famous hook company Partridge of Redditch and put me in touch with his very good friend Warren Duncan, while Hans de Groot introduced me to Al Worthington. Slowly, via handwritten letters, I started to build up a really meaningful exchange of correspondence with Al and Warren. It didn't take long before we were swapping flies, discussing our fishing techniques, and of course, sharing our experiences in catching Atlantic salmon on dry flies on both sides of the Atlantic. Al and I had developed a deep relationship although we have never had the opportunity to meet one another in person. He imported quality capes and saddles and was the owner of the Hackle Hut. I still have about forty letters from him, all written on his Hackle Hut notepaper, and some envelopes addressed to me from his flyfishing residence in Nova Scotia. All these amazing letters were filled with his latest flyfishing information, catch reports, and fishing techniques. Sadly, during a house renovation many years ago, I lost all my letters from Warren. But I consider myself very fortunate that I still have some of the flies that he specially tied for me to try on my trips to Nova Scotia, New Brunswick, Newfoundland and Labrador. I confess I have never tried them; they were simply too beautiful to fish with, so I copied them and have used the copies instead.

Warren was a great teacher and mentor to me, especially in perfecting my own hairwing patterns. I still marvel at his ability to create tiny little heads, even on bulky hairwing flies. Unfortunately, I only met Warren once and he died much too soon. I cannot adequately express my deepest gratitude for what Warren and Al taught me, and the information they shared freely with me.

In the footsteps of Lee Wulff

In my eagerness to learn as much as possible about dry fly fishing for salmon, I was most pleased with the personal correspondence I shared with Lee Wulff. We began somewhere in the mid 1980s with very technical correspondence in which we mostly discussed Lee's dry fly experiences and successes in Newfoundland, and mine in Norway. Lee realised that only a few people in Europe believed I had caught seatrout and Atlantic salmon on dry flies and I think this motivated him to help me. He sent me a few of his books and videotapes in return for some of my Klinkhåmer flies and Para Poly Sedges. I had included a handful of them in my first letter to him. For some reason, I doubted that he would reply to me. But he did, and he was even willing to share some of his flyfishing secrets with me, which greatly increased my respect for him. I had the pleasure of meeting him once, and the last time we corresponded by letter was in early 1990. Thanks to reading his books and watching several of his videos, my eagerness to visit Atlantic Canada grew to a fever pitch! Lee Wulff died on 28 April 1991 near Hancock, New York, when the Piper Super Cub he was piloting crashed into trees at the end of the runway during a landing. It was a big shock to me, but it also encouraged me to see the rivers that he loved so much. A few years after Lee's tragic death, I was able to make one of the most exciting flyfishing journeys imaginable. It took me several years of 5-6 week summer holidays to follow Lee's footsteps in western Newfoundland and on the great Northern Peninsula.

My way with Atlantic salmon

There hasn't been much written about dry fly fishing for Atlantic salmon in Europe, but in the few stories I was able to find and read I could clearly feel and sense that the majority of these articles were written from the heart. Most

A set of Bombers given to me by Marc Madore to try in Newfoundland and Labrador

Prefontaine Skate (left) and Whiskers variation are some of the flies Warren Duncan gave me

authors described their own particular flies, their own techniques and ideas, and that's why I would love to share my own vision with you: my thoughts, strategies and my flies as well. By doing this I feel you will get a good idea of how I do my flyfishing for Atlantic salmon, and how I have developed my own special dry fly patterns since the late 1970s.

One of the leading European experts with his own specific ideas on dry fly fishing for Atlantic salmon is Jesper Fohrmann of Denmark. Many may know him better as the 'fishmadman'. Another expert worth mentioning is 'Mr Pompero', whose real name is Heikki Anttonen. He is from Finland. A great help to me and a name that should not be missing here is Jim Gourlay from Nova Scotia, who introduced me to some of the best fishing guides in eastern Canada for catching Atlantic salmon on dry flies.

My first serious flyfishing attempt to catch Atlantic salmon was in 1972 when I was visiting the most northern part of Norway, and to be honest, I had no clue what I was doing. I had just turned 16 and it was my first holiday travelling by myself. I was just able to get my line out and had not even started fly-tying so the choice of flies was no priority at all for me in those days. I would pick any of my ordinary wingless and shoulder hackled dry flies and give it a try. I still remember the first Atlantic salmon I ever caught on a dry fly as if it were yesterday. The particular moment a large salmon's head breaks the surface just a few metres in front of you as he grabs your fly with barely a trace of his presence showing on the surface, is something incredible. Often salmon behave like this, and then sink down again, their great sickle tails just cutting the surface. It is an electrifying sight, one to quicken the heart of any flyfisher. I have seen flyfishers so astonished with what they saw that they simply forgot to set the hook. Head-and-tailing salmon not only keep your adrenalin up, but will test your mettle all the time. Moments like this you never forget.

I never wrote about the first Atlantic salmon I caught on fly, and I think it's an appropriate moment to mention it in this book. It all started after I ended up with a driver who was a fanatical fisherman while I was hitchhiking from Trondheim (Norway) up north. I even remember

Small lake feeders are ultimate hot spots for using dry flies for Atlantic salmon in Norway and Canada

Not easy to capture but beautiful to see, salmon leaping in perfect synchronisation at low water

that his first name was Svein. He didn't flyfish, but he was a great nature lover and he told me dozens of spectacular stories during the long trip from Narvik to Alta. It was a beautiful journey of just over 500 kilometres and as a high school student it was my longest hitchhike lift in Norway that particular year. Svein was unbelievably kind to me, paid for all the food, and we rested up en route with a close friend of his where we took a little nap. He even made a few detours just to show me the wonderful countryside. Remember, there is 24 hours of daylight at this latitude, and to be in the land of the midnight sun was like having a double holiday.

During our drive I told him about my addiction to camping in real wilderness and all the places I stayed during my way up north. On the last day he bought me a special fishing permit, some food, and dropped me off at the end of a small lake about 30 kilometres from Alta. It was an excellent place to put up a tent, and very close to the location where the river fed into the lake. He suggested that I stay there at least for a few days, promising me to pick me up again. Svein claimed that it was his favourite fishing hotspot in the entire area. He used to camp and fish there with his dad when he was still a boy. I was dropped in the right place at the right time, and a more picturesque spot to spend a few days in the wilderness was impossible to find. It was far better than any camping spot with all the luxuries, comforts and entertainment. It was August and the big run was on, salmon, seatrout and even sea-run Arctic char were showing off, leaping everywhere in the lake. Much closer they showed themselves in head and tail rises in the main current of the feeder stream. The water was low and the salmon were desperately waiting for rain so they could start their upstream journey to the spawning grounds. I couldn't believe there wasn't anybody fishing there.

It was an incredible fishing spot and I not only caught my first Atlantic salmon on dry fly, but also seatrout and lots of Arctic char as well. When camping out in the wilderness I don't mind eating some of my catches, and I can assure you that my breakfast, lunch and dinner were quite varied with all these species swimming around. But in the land of the midnight sun, life is a lot different in summertime. You eat when hungry and sleep when tired because you totally lose track of time.

Now back to fishing, and just imagine, I was only using simple wingless shoulder-hackled dry

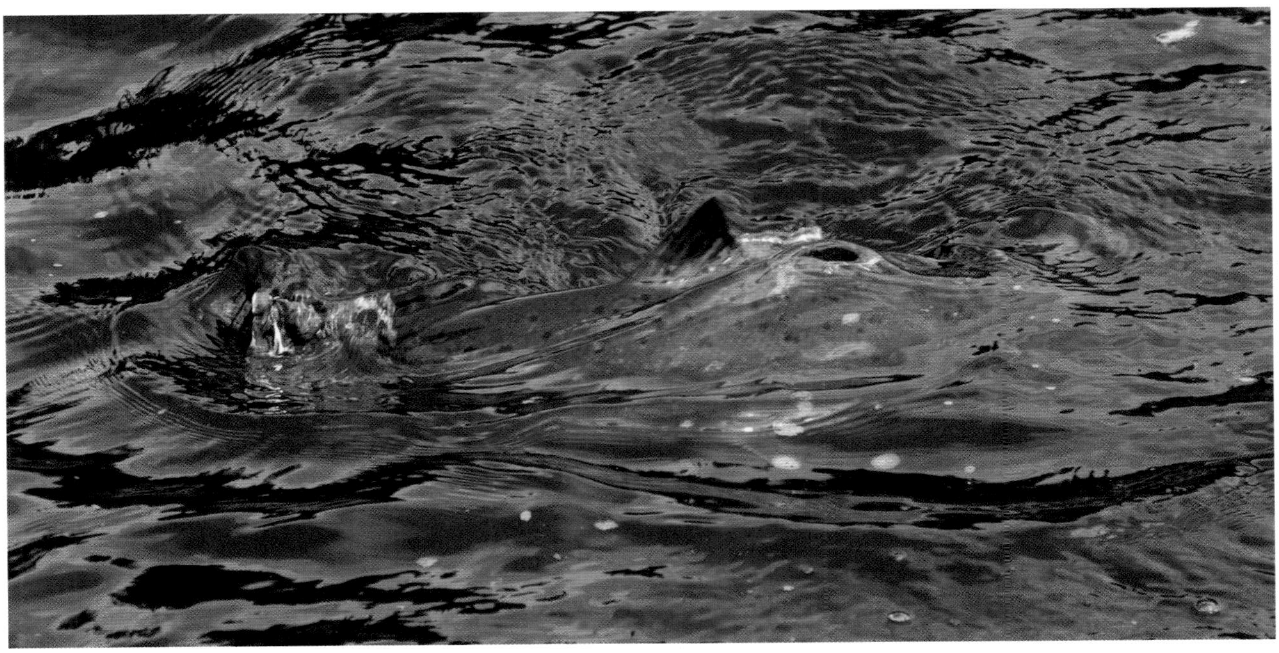

Salmon can take dry flies very aggressively as shown in this picture

flies such as March Browns, Pale Olive Duns, Large Dark Spinners, Brown Palmers, Red Tags, Bivisibles, and Cochybonddus and most of them worked. Big Bivisibles were my favourite by far, maybe because I had the most of them. Those days I didn't have wet flies or nymphs, so I was as pure as the best of purists. I didn't record my catches then, so I don't remember the exact numbers I caught, but catching four to six salmon (mainly grilse), a similar number of seatrout and at least a dozen sea-run Arctic char I considered an extremely good day's fishing. I lost many fish too, because my flyfishing skills were just one season old, but each fish I caught and landed gained me more and more experience. My casting was still awful and my fishing techniques very limited, but I was very proud and started to see myself as a real flyfisher already. I had caught a few real salmon too, probably in the 70-80cm range, all of which I released again.

When Svein picked me up a few days later I was not only full of exciting stories, but also was completely hooked on flyfishing for Atlantic salmon with dry flies, all the result of that one amazing experience. This fishing had put a spell on me! Amazingly, Svein believed me right away and explained to me that I had done extremely well, but he was amazed I had caught so many on dry flies because back then nobody was using dry flies for salmon and seatrout. Svein explained that he and his father only fished with worms decorated with red wool threads to add attraction for the salmon.

From my experiences in the 1980s and early 90s, I learned that as soon the air temperature gets higher than the water temperature, you can start using dry flies for Atlantic salmon. It is accepted as a rule of thumb by most of my friends, especially in Canada. I followed this line of thinking for quite some years and it worked pretty well, until 2015, when we caught a lot of fish on tiny dry flies in air temperatures far lower (3°C) than the river (9°C). Suddenly the rule had failed completely and was not reliable anymore. Exactly the same happened in the summer of 2016, and I wonder if this all has to do with global warming, or if it has always been like that and I had just missed seeing it. But I had kept diaries since 1984 and had never noticed it before. I also had my doubts about the weather being too warm, because in midsummer when the water temperatures gets much higher and the amount of oxygen has dropped to a critical level, most salmon will be hiding in the deepest and coldest lies with their noses down on the bottom, and they will avoid any activity to save their strength for their upstream journey to spawn. All you see is the white of the inside of their mouths as they open and close their mouths widely. This tells me that *now* is the time to leave them in peace. I don't like rules for salmon fishing, and I always keep in mind what Lee Wulff once told me: *'The only definite thing that you can say about salmon, is that you cannot say anything definite about them'*, and I really believe that strongly. I have caught salmon in a serious heatwave, but I have also discovered that the impact of catch-and-release can turn bad when water temperatures get very high. So I don't fish for them in these conditions anymore.

Rods for dry fly fishing

I started dry flyfishing for Atlantic salmon with 8- and 9-weight rods, but nowadays I mainly use 6-weight rods for low water and smaller rivers, and 7-weight rods for larger rivers and higher water. A powerful Thomas and Thomas Helix 6-weight is my favourite rod when using the smaller dry flies, while for big dry flies and wets I more frequently grab the same make of rod, but in 7-weight. Both rods have enough power to land salmon up to one metre in length. My 7-weight landed a 50-pound lake trout and many baby tarpon, so I trust the rod blindly. However, I take great care of my rods and they all go through a careful overhaul yearly. My 6-weight Helix landed lots of pacu in Malaysia, which means it's also an unbelievably powerful rod. Ina uses a Thomas

A set of large Bivisibles that I used frequently in my early days in northen Norway and Finland

In low water dry flies as small as size 16 can be deadly for catching sizes like this

Ina with a beauty that she caught on a size 14 Red Poly Bug with brown hackle as ribbing

and Thomas TNT 6- and 7-weight specially made for her with thinner grips, which suit her better. These rods are quite a bit lighter so she can fish easily for a whole day. She is very skilful when it comes to dry fly fishing for salmon, because that is what she does most of the time.

I must admit that I don't advise people to start with a 6-weight rod for salmon fishing because you really have to grow into that. For grilse a powerful 6-weight is fine, but before using it for bigger salmon you should have first gained some experience. I don't like long plays when fighting salmon, or any other fish, and although I haven't studied the effects of playing different species of fish, it simply doesn't feel right. Some people say playing a fish for too long will kill salmon, but I have seen many surviving unbelievable damage when they have smashed themselves on the rocks over and over again while trying to jump the waterfalls they encounter on their way upstream. Of course, several do kill themselves this way. In 2018 Ina even caught a salmon that took her dry fly and was so badly damaged by a seal or an osprey that you would have thought it wouldn't come up for a dry fly any longer. But it did, and it was not the first. I have also seen Atlantic salmon spawning despite serious bite marks from seals, and caught fish with big nasty wounds. I also helped out during a research project by tagging fish that I caught and released, and they were found later back at their spawning areas. Most criticism of catch-and-release fishing comes from people who want to keep fish because they are worried that a fish will be too wise to be caught again after it has been released.

I can categorically tell you that this is nonsense, because I have caught salmon with one or even two flies in their mouths or their bodies and they still grabbed my fly. Some people claim only 10% of Atlantic salmon take a fly well, and those takers still take flies over and over again after being released. But I have never dug deeply into it. What I did discover about playing salmon is that when you fish with light tackle, salmon leap a lot, which makes them tire more quickly, and I can land them not only faster, but my fishing partner can get a lot of photographs of salmon leaping high into the air as well.

Leaders

In general, I am not the best long distance caster and therefore my strategy is to use the wind, the current, unusual fishing techniques, and special, well-thought-out flies as my most powerful weapons. Self-made braided leaders help me a lot in the presentation of my dry flies, and when I use big dries I even use a different leader-tippet system. For me the backing, fly line, leader and tippet are four totally independent parts of my fishing system, although I connect them in perfect harmony, and hardly use any knots. The only knots in my entire system are the backing to my fly reel; then comes the leader-tippet construction and finally my tippet to fly connection. I developed this unusual system in the early 1980s, together with the late Leon Chandler, after I had lost a big salmon and a grilse on the same day, clearly due to bad backing and leader connections. A big salmon had taken my entire fly line when it broke off at the backing in its fight for freedom. And my whole leader went with a grilse just two hours later. (My fly line was found the next day without fish.)

For almost a year I put all my efforts into finding a system with as few knots as possible, and together with Leon's help, I accomplished this. Since 1982 I have never had a similar problem again. Years later, when I started to do more exotic saltwater flyfishing, I realised even more how important this system really is to me. Of course, I lost many flies and broke off lots of tippets, but since then my leader, fly line and backing connections have always remained intact. Leon was a very close friend, a real gentleman, and one of the most modest flyfishers that I ever met. He was widely hailed as the 'flyfishing ambassador of the world'. Before his retirement he was vice president of the Cortland Line Company, where he had spent 51 years of his life. In the 1980s and 90s Leon and I shared the most useful correspondence I've enjoyed with any of my friends. Most of our topics were about fly lines, tippets, flies, leaders and backing materials, and I feel very privileged that I was able to meet him several times and that we have been such close friends for so many years.

I adapt my tippet length to the weather and water conditions, and when there is no wind I also don't use very long leaders because I love to use the wind to benefit my casts. I learned this style in the highlands of Norway where I had to deal with unpredictable mountain winds all the time. My leaders go down from 30-20 to 10lb and all parts are connected without any knots as well. The biggest benefit after developing this backing, fly line, leader, tippet system is the unbelievable success when casting my big Klinkhåmers and the surprisingly accurate presentations that I am able to achieve. The braided leader reduces wind knots to a minimum while at the same time letting my large parachute flies or big Bugs and Bombers turn over very smoothly.

My braided leaders work extremely well in combination with the light, 6-weight 10ft Thomas and Thomas Switch rod but also with the Thomas and Thomas TNT and Helix single-handed rods in weight 6 and 7. I use the Switch rod mainly for my ladies' and kids' classes which makes it easier for them to cast a little further when necessary or when they need to make a roll cast. Because of the braided system I have very few wind knots and when a tippet breaks off it's very easy for the kids to connect a new one. For my classes I have added a loop to the butt section for quick rigging. For myself I connect the leader straight to the fly line again without using any knots. I create a loop or I connect a strong tippet ring at the end so the leader will last a whole season if treated well.

Tippet talk

Salmon in general are not tippet shy when fishing for them with wet flies, or even when using wet flies tied on with a Newfoundland or Riffling hitch, but when fishing dry flies in high summer and low water conditions, things can be completely different. My choice of tippet depends a lot on the

Beautiful salmon caught with an orange size 12 Poly Bug in Fischell's river, Newfoundland

surface of the water and the weather conditions. In calm water, smaller dry flies and thinner tippets are much more effective, while during rainy and windy conditions, or when fishing on disturbed surface currents, bigger flies are absolutely essential. For example: in 2016 I caught two nice grilse on a size 14 dry fly in the same pool in almost windless conditions. Forty minutes later I landed a 89cm salmon on a size 16 dry fly, and two hours after that I hooked and landed a 96cm salmon on a size 1/0 Bomber in heavy rain. Salmon can be unbelievably active on dry flies in pouring rain, and my experience has taught me to go for big flies under these conditions. A lot of anglers remain extremely loyal to the same tactics, flies, fly sizes and techniques and use them year after year. I always memorise tactics that worked, but I also experiment a lot and keep updating my knowledge all the time. I do this by using my fishing diaries, where I record my flyfishing techniques, the flies I used, the prevailing weather conditions and my catch records.

I often try something completely different when success eludes me, but I also try different things when the catches are plentiful. This is how I discovered that nymphing for Atlantic salmon works really well, and I learned that big salmon do indeed take very small dry flies, much more readily than we often imagine. When Lee Wulff told me that he landed quite a number of salmon with a 5-weight rod on a size 16 dry fly, I was not that surprised and believed him right away, because I had caught dozens of salmon on size 12 and 14 dry flies in Norway in the 1970s and 1980s myself.

Many people must have been very sceptical when he spread that story, and I know exactly how that feels. Maybe that's another reason why Lee and I felt so close to one another, and why we had such fruitful correspondence over so many years. With salmon fishing I like to do things that are different to what most of the locals do. So when everybody is using an orange bug I will absolutely grab a black one, or even a purple one, or just tie on a Klinkhåmer, a Rackelhanen or Para Poly Sedge. I will do exactly the same with the sizes of my flies and when most people are using a size 6 Bomber I will go down to sizes 10, 12, 14, or do the opposite and use a size 2 or even size 1 fly. It is a great method that has worked extremely well for me, and this is perhaps one of the main reasons why I haven't had a blank trip for Atlantic salmon since 1988. Being immensely creative when fishing for salmon is very important.

Every flyfisher knows that small flies need thin tippets to give the fly a better natural drift, and a size 12-14, or especially a size 16 dry fly, will not float or drift naturally when tied on a 0.30 or 0.32mm tippet. Believe me, for dry fly fishing in smooth-surfaced waters, the diameter of the tippet is crucial to succeed. I have proven this many times while testing the best of the best monofilaments, trying to find the perfect balance between the smallest diameter of tippet material and largest possible size of fish I can catch on it. When dry fly fishing, I always search for the spool with the strongest breaking strain and the smallest diameter. I prefer Stroft GTM, LS, or ABR, because I know from the great number of practical tests we did, that I can go almost 0.1mm thinner compared to most other brands, and that is essential when using very small flies as I so often do nowadays. The difference between using a tippet of 0.30mm or just 0.20mm is catching fish or not. I know from experience in Norway and in Atlantic Canada, that when the average size of the salmon is not more than 85cm (approximately 6kg), I can safely use a minimum of 0.20mm Stroft GTM, while most other anglers are still using 0.30 or even 0.35mm from another brand. However, most of the time I use a 0.20-0.22mm tippet for my dry fly fishing, except when there are a lot of rocks and boulders in the water when I go to 0.24 or 0.26mm. I never go thicker than 0.26mm, and this is the same tippet diameter that I use in my classes for the kids and women.

After testing countless brands of monofilament, Stroft turned out the best for me by far

Rising fish

Because of my poor casting in my early days of flyfishing, I was lucky that Atlantic salmon can be curious and very aggressive at times, which helped me to succeed. Over a period of 53 years dry fly fishing for salmon I learned a lot from good observation and while some salmon will come up for your fly once, others will do it twice, or even five or ten times. Many times they will just nose the fly, or push it away, or even try to sink it with their bodies, especially with the tail.

Sometimes you know exactly when to change your fly, such as when a salmon is in the mood to come up on almost every cast. Of course, it depends which part of the world you are fishing in. It also depends how often you change flies, because dry fly fishing for salmon means changing flies a lot. I have even experienced several occasions when a salmon has come up only once, each time you changed your dry fly, but yet never actually taken it. I have also experienced situations when I raised a fish with dozens of different flies without hooking it and then Ina hooked that particular fish with her first cast. If I tell you that we have seen the same salmon come up more than thirty times for a fly in Newfoundland, you'd be forgiven for calling me a liar. Grilse can come up even more than that! Important to know is that I have also caught quite a good number of salmon on dry flies by just irritating the fish so much that they finally grab the fly. It only works, though, when the salmon are head-and-tailing in exactly the same spot, and when they do, I just let the fly splash down in front of their noses until they finally grab the fly. This technique works best in Norway for me.

About 80% of my dry fly fishing for salmon is done by sight. I learned this in the northern parts of Norway and on the Fosen Peninsula, northwest of Trondheim. The Fosen Peninsula is one of Norway's best areas for dry fly fishing as I discovered in the 1980s and early 90s. On this wonderful peninsula I'd found a particular spot in saltwater, a few hundred metres long, which narrows so that it flows like a river with each changing tide, and flowing in a different

For Ina fishing for Atlantic salmon always means dry fly fishing

Another nice fish that grabbed a pink Poly Bug this time

direction with the change of tides. It is one of the few spots in the world I know where I am able to catch salmon on sight in saltwater with dry flies. Not many, but at least a few every season, when we spend two weeks in this wonderful fishing paradise. Once, in prime time, we caught seatrout by the dozen in the same spot during one tide, and all on dry flies in sizes 14 and 12. The seatrout we were after were all head-and-tailing fish, and I've been fishing here for at least ten years, in the good company of a few of my closest friends. One of the best dry fly rivers that I ever fished for Atlantic salmon in central Norway was the beautiful Nordelva in Fosen. I averaged a minimum of eight fish a day until they built a road to make the river accessible to fishermen and the wonderful dry fly fishing was over. Norwegian salmon just don't like to come up and grab dry flies when the riverbanks are crowded. I've had the same experience when fishermen using wet flies put their drifts into very smooth water over and over again. They spoil the fishing for a dry fly fisher. For dry fly fishing I prefer to be totally alone.

Studying the behaviour of salmon taking insects. Here a salmon grabs an ant that I threw in the water

My best spot in the world for learning all about Atlantic salmon and their eager taking of dry flies

This is what happens when I throw in a handful of forest ants. Believe me they really grab them

Dry fly observations

Have you ever seen salmon taking dry flies from the surface? I have, and I still see it every year when I visit my favourite rivers in Newfoundland and Labrador. I have also seen salmon taking dry flies in Norway, especially in the most northern part and on the Fosen Peninsula. However, most people will tell you that salmon do not feed in freshwater because that is what we learned at school and from our scientists.

Personally, I don't know if they really feed or not, but I know they take dry flies and insects off the surface. But it happens irregularly and not in every river system. There is also a scientific report that proves that in the Tana River system in northern Norway, the stomachs of 45 out of 133 adult salmon in the River Laksjohka (a tributary of the Tana) contained food items. I have also found insects (often bees) in salmon stomachs. This is still not clear proof that Atlantic salmon really feed, but it does prove that they grab flies or other insects from the surface and swallow them. This might be also be the reason why I was catching so many salmon on dry flies in the northern part of Norway in the 1970s and early 80s. I have heard of insects found in salmon stomachs in Atlantic Canada, but not as often as I have heard of it in Norway. Why do they grab dry flies so keenly? The answer to me is very simple: it is their feeding memory from when they grow up in the river and feed on mayflies and other insects. A head-and-tailing fish coming up at the same spot over and over again is for me a target for dry fly fishing, but so also is a salmon playing with foam bubbles on the surface, as are those salmon that seem to lift a little when you drop a bomber or bug on the water.

There are many theories about their feeding memories, or their feeding behaviour, but I keep it simple because I am not a scientist. I have seen salmon take insects from the surface many times and that is a fact. However, whether every salmon that grabs a real fly from the surface also swallows it, I do not know.

Tying my first dry flies for Atlantic salmon

Once my flies started to look better I became more and more confident, and finally became a real fanatic, and a creative tyer too. After a while I was so highly motivated that I was tying flies almost every day. Of course, like anybody else, I started with simple flies, such as a Red Tag, Bi-visible and Palmers, but as soon as I knew I would spend more summers in Norway, I started to tie more visible, bigger and better floating flies for salmon. I increased the number of hackle feathers and used far more wraps, or made the bodies a lot thicker and created different body shapes by using long saddle hackles that people gave me in the USA. And believe me, I experimented a lot, including tying patterns with and without tails.

I often skipped the vital proportions and took to tying a lot more creatively. I had saved a lot of old flies that had been almost completely destroyed by salmon, Arctic char or seatrout, and I used these as blueprints to develop my own patterns. Each fly I finished I dropped in Permaflote, and as soon they dried, I tested them in a little dish of water. To begin with I used all sizes of hackles, but as soon as I started tying my own bugs and bombers for Atlantic salmon, saddle hackle soon became my favourite hackle material.

Polypropylene yarn again

The first big change in my dry flies for Atlantic salmon came somewhere in the early 1980s when I started tying the first in a series of brand new flies based on the experience I had gained from tying and fishing the Rackelhanen in previous years. I used poly yarn that I bought a lot in Sweden to tie my own Rackelhanen variations, but when I made my first salmon flies with polypropylene yarn there was a huge difference. I was extremely lucky when a good fishing friend of mine, who knew my preference for poly yarn, found a lot of it in many different colours in

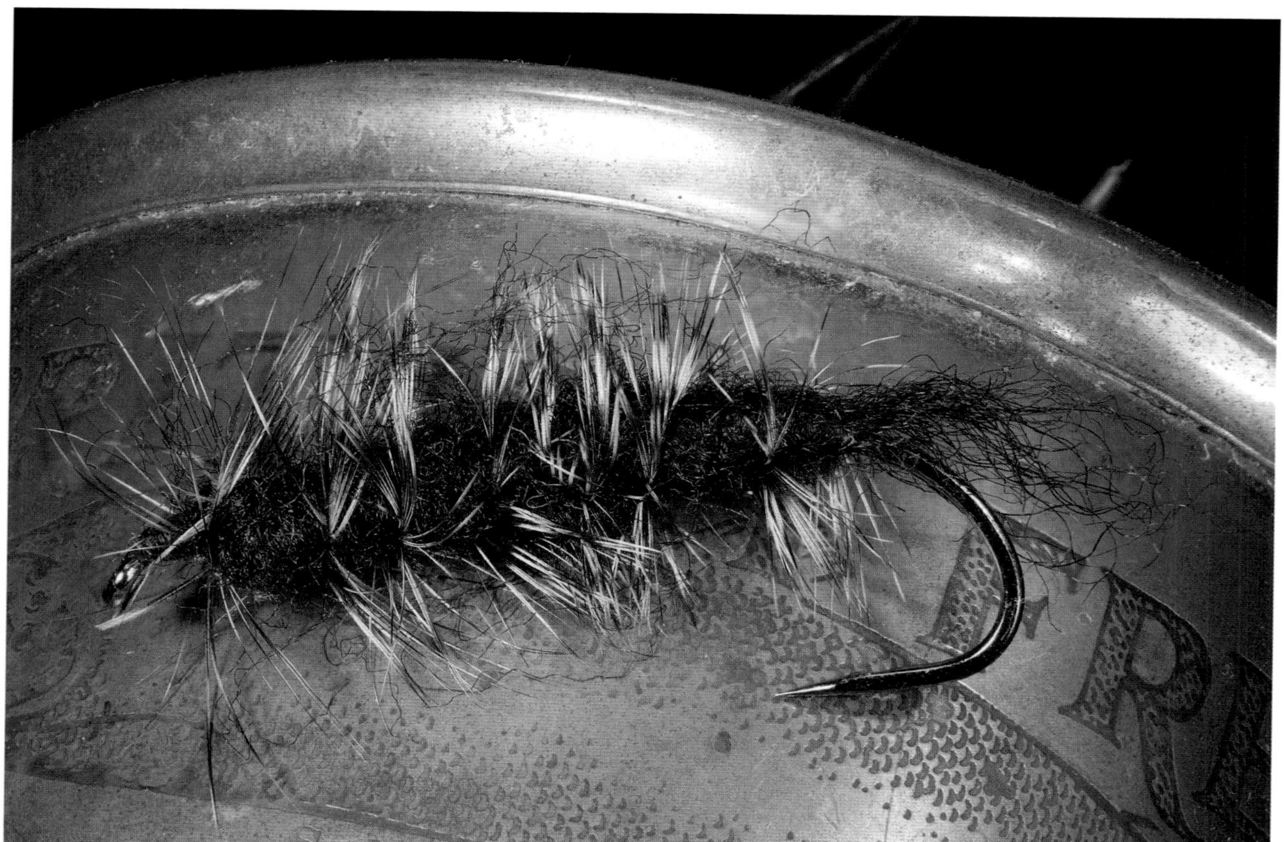

My first Poly Bugs for Atlantic salmon were pretty big

A salmon takes a Green Poly Bug

Italy. The yarn was wrapped on big spools and he shared at least ten big spools with me, each in a different colour. A lifetime supply for sure! It then became really great fun to experiment with this awesome synthetic material that was suddenly so abundantly availably in my early days of flytying. For the first time I was able to work with poly yarn as body and wing material unlimited, and later I discovered a bit of magic. The magic was actually making my own poly yarn dubbing by cutting off small pieces and putting them into a coffee blender. Besides using plain colours I also mixed my own shades. Using the experience I gained tying the Rackelhanen flies, I found that a combination of Permaflote, poly yarn and hackles turned out to be perfect to tie floating dry flies for salmon fishing. The flies looked quite ugly, because they had shaggy, dubbed bodies, but I was so happy with the results, and the way these patterns floated in my dish of water, that I had great expectations for the future. From below, the flies' silhouettes looked extremely attractive and their bushy bodies made them very buggy looking. A great attraction for fish.

I love working with poly yarn, and it's one of the materials in my tying I use most and has been since the early 1980s, more so after I was able to source my own spools of poly yarn.

Poly yarn tied on a hook has to be treated with a high quality floatant or the yarn will immediately absorb water and the fly will sink under the weight of the hook alone. However, the better news is that while poly yarn absorbs water well, it also absorbs floatant really well too, and floatant actually makes it water-resistant, even waterproof. Many people don't seem to know this. My floatant today is Dilly Wax from Fly-Rite, which is specially made for extra fine poly yarn dubbing. It's non-toxic and can even be used as a lip balm.

I make my own dubbing from poly yarn for my large sedge patterns, such as the Rackelhanen, and Para Poly Sedges and for most of my special dry flies for salmon. For all my other dry flies with synthetic bodies I use Fly-Rite extra-fine Poly-dubbing material. Because many materials are sold as poly yarn, it is wise to do my special floatability test as I have described so clearly in previous chapters.

Hooks

I have done extensive research on hooks that I could use for dry fly fishing for Atlantic salmon. Sadly, a lot of the hooks that I liked in the past have been changed from the original design, or are not available anymore. Since I use the Klinkhåmer Special for my salmon fishing a lot, I prefer the Daiichi 1160 and 1167 Klinkhåmer hooks, but only in sizes 8 and 10. Personally, I also use Klinkhåmer hooks in size 6 and sometimes even size 4, but these two sizes are sadly not commercially available and belong to my private collection. Besides hooks with a straight-eye, I also use hooks with a down-eye for my salmon

This beauty straight from the ocean took a larger Blue Poly Bug (size 4) in the rain

This fellow couldn't resist a yellow Poly Bug tied on a size 8 hook

flies, dry or wet, and when more and more down-eye salmon hooks disappeared from the market, I developed my own salmon hook a few years ago. Today the hooks are known as the Daiichi model 2110 (bronze) and 2117 (black nickel) Bomber hooks. I wanted a hook that I could use for bugs, Bombers, skaters, Wulffs, Walkers and of course for my very special Bondal series of wet flies as well. The hooks are very popular in Atlantic Canada and are also popular for tying salmon flies for the Kola Peninsula, but are lately commonly used for tying hoppers, crickets, and other large terrestrials. I even use size 10 to tie my Leadhead version 3.0.

For my special and very small Poly Bugs, which I tie from size 6 down to size 16, I use another hook I sourced privately that sadly is not available on the market and also was a special gift to me from Alan Bramley. These are wet fly hooks.

How I get my private hooks is easy to explain, in that time when there was a 'failed' production, or if a customer wasn't fully satisfied with the shape, wire or the eye of a certain hook, Alan gave them to me to use in my workshops and classes. He gave me many, and I have only a few hundreds of these kinds of hooks left right now. The good news is that Daiichi have two excellent substitutes in the models 1530 and 1550. These hooks are as good as my private hooks; both have sproat bends, down-eyes, and are not too long in the shank, and most importantly, they are extremely strong, which makes them ideal for tying little bugs which are unbelievably popular in Atlantic Canada. The 1530 model has an extra heavy wire, while the 1550 has a standard wire hook. My private hook is standard wire too.

For some of the Poly Bugs I tie for very special friends, I only use the 1550, and so far, everybody is very satisfied with them. For the smaller sizes, however, I prefer the 1530, just in case a real whopper grabs the fly. I also use the 1550 for my kid classes and many children tied their first ever salmon fly with it.

For all my salmon dry flies that need a longer shank, such as my Para Poly Sedge, I prefer the Daiichi 1760, in sizes 10 up to 4. These hooks are specially designed for tying steelhead flies. For normal size Bombers or Wulffs I highly recommend the Daiichi model 2110 or 2117 in size 10-4. Good quality and big dry fly hooks for salmon or steelhead are very sparse. For my bigger Bombers and dry flies for salmon I use the steel blue Alec Jackson hook but honestly,

My set of widely available dry fly hooks

Some prototypes of experimental bugs from the early days

I prefer a similar hook without any curve in hook shank and with straight eye. If you like it you can also tie big dries on tubes like Jesper Fohrmann does.

Deer hair versus poly yarn

Thanks to poly yarn I developed quite a few unusual dry flies for Atlantic salmon at an early stage in my salmon fishing, and my catches increased a lot. I ascribe the pattern's successes to its high floatability, especially when I compare them with most of my previous salmon flies and deer hair bugs. The newer patterns shared the fact that they were based on the body material of poly yarn dubbing and the cigar-shaped body of North American bugs and Bombers. So what I actually did was change the body shape of the Rackelhanen and added turns of hackle around it. When I first left the wing of the Rackelhanen out and added a tail, the flies started to look totally different.

Later, when I became more familiar with the dry flies for salmon, especially for salmon in Atlantic Canada, I realised that I hadn't developed anything special, new, or different at all, but had simply created my own kind of salmon bug, very similar to those used in Newfoundland and Labrador, New Brunswick and Nova Scotia, but using a completely different body material. The main difference was that I used poly yarn as a body while the Americans and Canadians use deer hair for their bodies.

The most successful bugs I tie with deer hair are my bumblebee imitations, and in many different colours. I started fishing them a lot after I noticed a salmon taking a wasp that was floating in the Salmon River at the northern tip in Newfoundland. This was in 1999, during a period of drought and extremely low water conditions. There was, in fact, a small heatwave on the go in Newfoundland, and the lack of wind over a long period made it all the worse. They were not the best conditions for fishing, and we didn't see any other anglers at all. Our catches were poor so I was concentrating more on the fish we spotted in the river than I was on the actual fishing, when I suddenly saw this salmon grab a wasp.

That same evening, I started tying some wasps using yellow and black deer hair. The first ones were far too big because I just repeated the usual size flies we were using in those days, like sizes 6 and 4. But then I had caught a real wasp and I made a better copy of its size and I had my first success fishing the new pattern. It soon became apparent that a flawless copy was not needed, just an accurate imitation of *size*. That meant a good cigar shape and, of course, some colour match. So, I tied them on size 8 and 10 hooks and hackled them with a brown saddle hackle because black was hard to see in the dark brown waters of the Salmon River. Since then, for many years I have tied at least 30 different colour variations and each year we have had our favourite colours. This experience of tying and fishing with deer hair paralleled my Poly Bug development and it was very important to get the colours right, as it was in my Poly Bugs as well. It took a while before I could call them the Perfect Poly Bug.

To be honest, I don't like tying deer hair flies because deer hair fibres end up all over the house. That's why my good friend and professional flytyer, Gordy MacKinnon, from Sydney Mines (Nova Scotia), makes them for me and has done so for the last ten years, and in a huge variety of colours. I supply him the correct hooks and tell him the sizes and colours I want. In return I give him material for his kids' programmes and thus we help each other.

More time spent to perfect my Poly Bug

After testing and improving my Poly Bug, it still took seven years before I was completely satisfied. It's so weird that one of the easiest and most simple flies to make took me such a long

Ina's box of Bumblebees with great colour variations to be sure of success (All tied by Gordy)

Gordy MacKinnon, posing with one of the many sets Bumblebees that he made for Ina and me

Nice grilse took a pinkish red fluorescent Bumblebee

time. I constantly had to compare my special creations with those tiny deer hair bugs as well. Deer hair bugs were excellent to fish with and you will catch a lot of fish on them, but I was worried by the fact that salmon nose them a lot, or just play with them, instead of taking them properly. It's unbelievably exciting to see salmon coming up to deer hair bugs, but we missed too many fish when setting the hook. Through a serious analysis and evaluation of both patterns, and by being creative with hooks and materials, I was eventually able to arrive at the final version of my Poly Bug.

After many years I had at last found a perfect harmony between all the materials and colours I was testing. And I found that my homemade poly yarn dubbing has several benefits over deer hair for my Perfect Poly Bugs:

1. You don't need to be an experienced flytyer to tie this fly and can make it in two minutes.
2. I can mix my dubbing easily and make every shade of body colour I want.
3. The flies are almost unsinkable if treated before fishing with Dilly Wax.
4. The rough poly yarn fibres create a better silhouette and they also trap air bubbles.
5. The hackle used as ribbing merges deep into the poly dubbing body which makes it a very strong and durable construction.
6. The Depth Ray fluorescent nylon wool is not only extremely durable, but more importantly, it holds its original colour even after many days of fishing.
7. The more fish caught, the uglier the fly gets, but the more fish are keen to take it.
8. I simply named the flies after the various colours of the Nylon Translucent ribbing material, and that is exactly how I chose the names such as: Green Poly Bug, Orange Poly Bug, Pink Poly Bug, White Poly Bug and so on. The colour of this ribbing is the only clear visual way to recognise the differences within the flies in my series.
9. The main body dubbing is just black for this whole series, but of course I tie variations with other dubbing colours as well. Poly Bugs with bodies in purple, orange, peach, and even magenta will work very well under certain conditions too. The colour variations are unlimited. Ribbing colours can also be changed and here you can test your own creativity and inventiveness in time.
10. Thus far I haven't found anything wrong with my final Poly Bug creations, but the only reservation I have is that because I am using black poly yarn dubbing in combination with a black hackle, the fly is quite hard to spot in waters that contain tannic acid. There are many tannin-stained rivers in Atlantic Canada and even several in Norway too. People who are not used to fishing these clear but dark brownish coloured rivers might have some serious problems spotting the black hackled versions. I actually made it for reflecting bright, shiny water, but then found that this particular colour combination seems to be the best of all under any circumstances, even in rainy conditions. When I can't see my Poly Bug, or I suddenly lose track of it, I mostly set the hook on any swirl or rise I see, because a lot of salmon seem to love swallowing the Poly Bug off the surface. To make the fly more visible you simply add a brown or ginger hackle and suddenly it's very easy to spot. I do this for kids and ladies all the time when I have my dry fly salmon classes and workshops.

With light tackle and small dry flies, hooked salmon can easily leap up to 15 times

Salmon with a yellow Bumblebee with brown hackle for better visibility

This whopper took a size 16 orange Poly Bug while I was fishing for brook trout with my 5 weight T&T rod

Complete series tied in nine colours to create nine different Poly Bugs

The Green Poly Bug with black hackle is hard to see but the most successful Poly Bug so far

Important notes about the materials in the Poly Bug dressing

1. My Perfect Poly Bugs need to be tied shaggy and most of all ugly. They must look more like used flies to be most effective. If you tie them too beautifully, they are less effective. The shaggy silhouette is the big difference against the more beautiful deer hair bugs, and gives this fly the extra power that it needs.
2. For early season and normal weather conditions and colder water I use sizes 6 to 10. During rain you can go to size 4 and 2, but once temperatures start rising, and for windless days, I slowly drop from size 10 to size 16.
3. Daiichi hook 1530 is a little stronger and I use it for the smaller sizes because the chances of hooking bigger salmon in these sizes are fairly good in Newfoundland and Labrador.
4. GSP threads, of course, will work, but most are quite slippery and there is not a nice harmony between GSP and my rough homemade poly yarn dubbing. This is why I recommend using threads like Semperfli Pure Silk or Semperfli's Classic waxed thread. Even with waxed thread I still use my very sticky Overton's wax to be sure my poly yarn dubbing will stick well onto my thread. It makes the twisting of the dubbing a lot easier. You need to twist the poly yarn dubbing very tight which is why you need at least 6/0 thread. For myself this is essential because I use a lot dubbing and use exclusively an electronic dubbing twister, which makes my life a lot easier and twists lot faster too.
5. Believe me I have tried many materials for the tails. Black deer hair worked quite well, but it isn't very durable. Calf tail sadly gets waterlogged more quickly but I like it the most, so I always use a bit of extra Dilly wax on my tails. Calf tail as main tail material means the bugs always float in the correct position and create an awesome silhouette too. Flies tied with calf tail for the tails always caught most fish. Please note: For the smaller sizes (12-16) I use calf body hair instead.
6. As I mentioned, I make my own dubbing for my most perfect Poly Bugs. I start by folding the poly yarn strands in several layers and then start cutting 0.5cm strips from them which I collect in a little cup. As soon I have enough I put them in my coffee grinder and make enough dubbing to tie at least a few dozen patterns. Sometimes I work ahead and just make dubbing, which I store by colour in small plastic bags. **The dubbing together with the special ribbing material are the most essential materials for making this peculiar fly so effective.**
7. For the main ribbing, I actually use Depth Ray fluorescent nylon wool from Danville. This is for several reasons. I have used it exclusively as tags for my Bondal series since 1988, so I am very used to this material, and it has been tested for years. It's extremely durable and colour proof. The variety of colours allows you to tie several different flies within this series. I use the wool in nine different fluorescent colours, namely: two variations of red, yellow, orange, green, fire orange, white, blue and pink. Each colour represents an individual fly in my series.
8. Making this special series works extremely well, especially when I have to change patterns according to the behaviour and taking mood of the fish. When salmon come up a few times for one particular colour, but still refuse to take the fly, you change the fly to a different colour. I do exactly the same for a fish that has a look at a fly several times, but then stays down and refuse to come up again. When you change to another Poly Bug with another colour ribbing, you often can trigger the same fish again. This is one of the big discoveries after testing my Poly Bugs carefully for the last seven years.

Ina's first salmon caught with an orange Poly Bug

Master guide and close friend Junior Flynn holding my rod after this beautiful catch with a Green Poly Bug

Mr Pompero and me posing at the Fly Fishing show in Tampere Finland in 2022

Tying the Green Poly Bug

To tie another colour Poly Bug you only need to change the ribbing colour of Danville's Depth Ray fluorescent nylon wool. I have picked the Green Poly Bug because it is my most successful in the series.

Dressing: Green Poly Bug

Hook: Daiichi 1550 size 4-6-8-10 for early season and Daiichi 1530 size 12-14-16 for high summer

Thread: Semperfli 6/0 Classic waxed thread, black

Tail: Black calf tail or calf body hair about 15mm long. Max 20mm

Body: Black homemade dubbing made from poly yarn strands mixed in a coffee blender

Ribbing: Danville's Depth Ray fluorescent nylon wool in green colour

Hackle: My preference is a black saddle hackle but to make the fly more visible when I tie this pattern for other people I use other colours like brown and ginger too

Tying steps

1 If you read my chapter about the Rackelhanen you know exactly how to make your homemade dubbing from poly yarn! For my special Poly Bugs I use black dubbing.

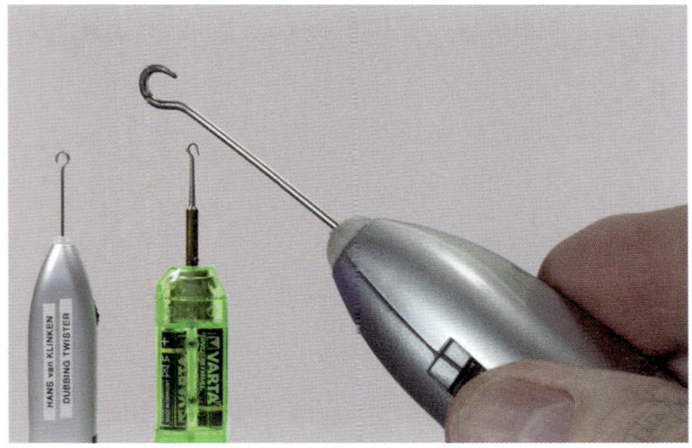

2 When the dubbing is made, I test and eventually adjust my electronic dubbing twister so it works very smoothly. I usually made a lot dubbing in several colours before I start tying!

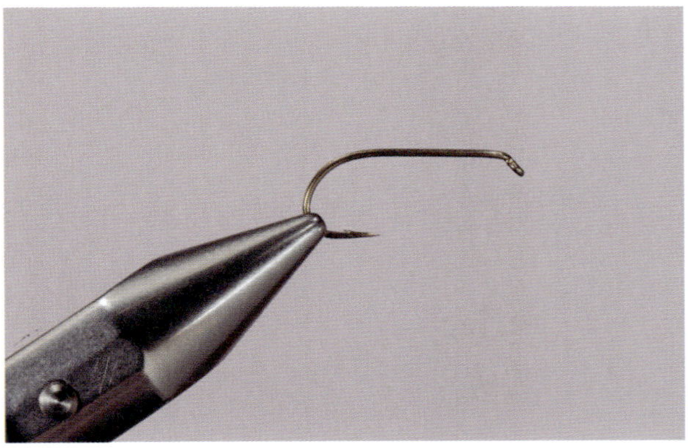

3 Squeeze the barb and place the hook in the vice.

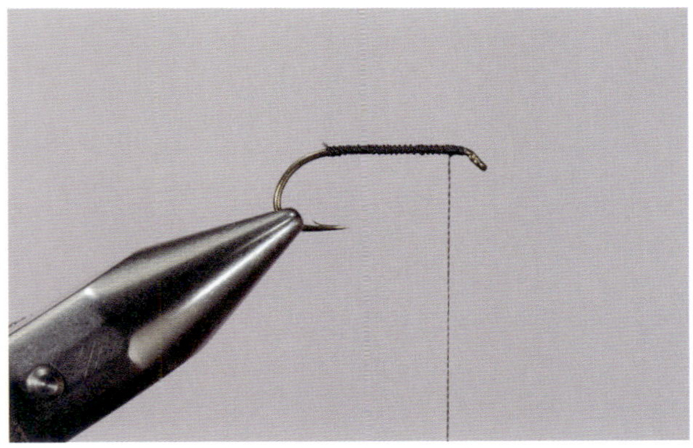

4 Cover the entire hook shank with closely spaced wraps of thread, starting from the eye.

5 Tie in a little bunch of black calf tail as tail and secure extremely well because you don't want the body to twist around the hook shank later.

6 Now you tie in a black saddle hackle by its tip.

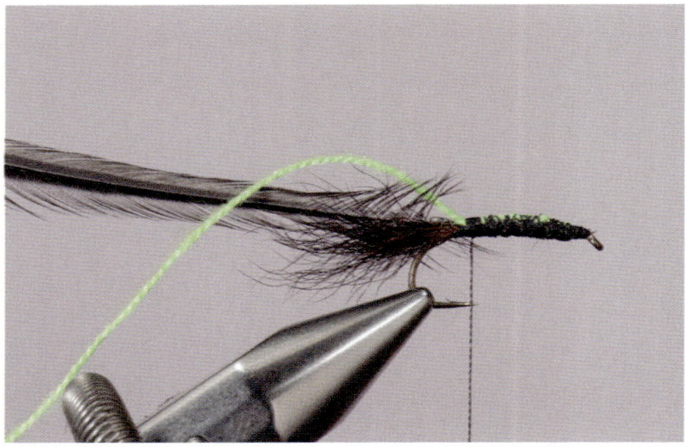

7 Tie in the Depth Ray fluorescent nylon wool in green colour for ribbing.

8 Lengthen your tying thread and cover it well with high quality wax.

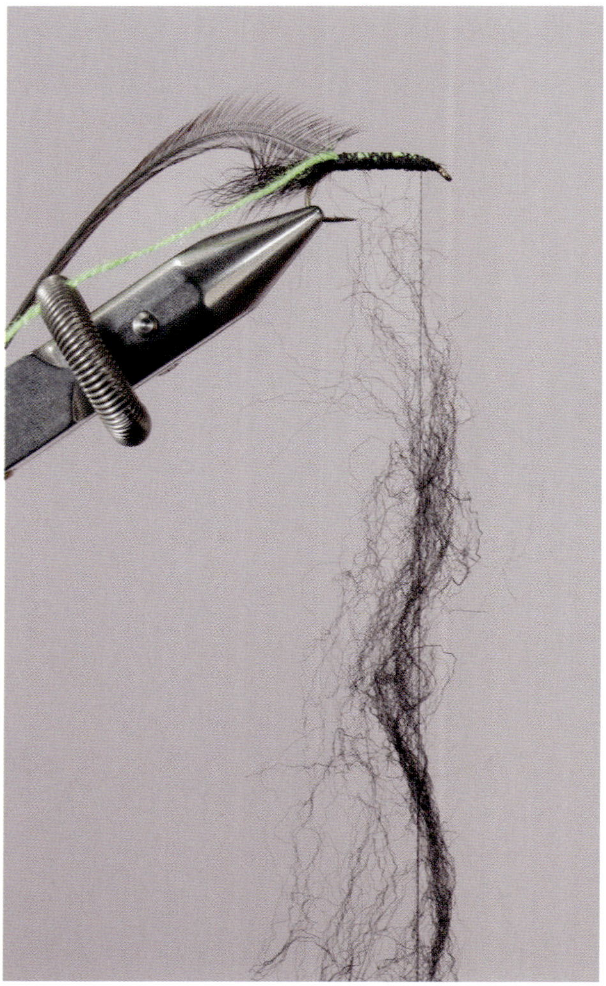

9 Put some dubbing on the extra waxed thread. Don't worry if it looks a bit shaggy.

10 Squeeze the dubbing under some pressure from thumb and point finger, and twist it in one direction until you have covered approximately 20cm of thread with dubbing material.

11 Make a loop and secure the thread to the hook shank and bring the thread forward.

12 Put your dubbing twister in the loop and make a very tight dubbing loop. As you can see my simple electronic one really works very pleasantly under almost every angle.

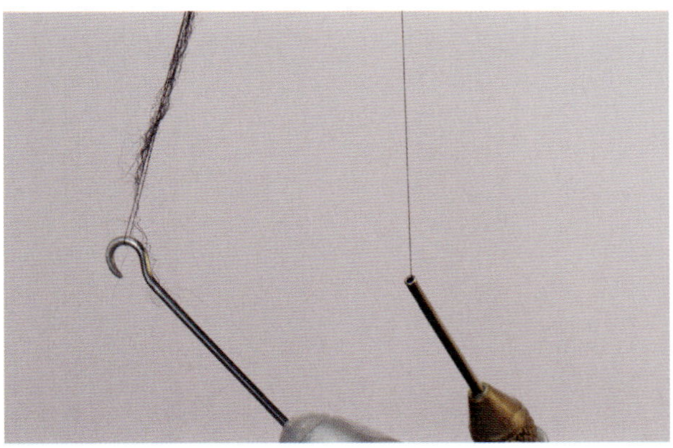

13 Close up: See how easily I catch the loop with my twister hook.

14 Dub a nice cigar shaped body, secure your thread and cut off all waste.

15 Use a dubbing brush to make the body shaggier.

16 Grab the nylon ribbing wool and wrap it forward in opposite direction.

17 Secure the ribbing and cut off waste.

18 Take the hackle and wrap it equally with the ribbing.

19 Give the wraps a bit of tension so the hackle quill gets pulled a little deeper into the body and make the fly more durable. Cut off excess hackle part.

20 Make several extra windings and tie off with your whip finisher.

21 Completed fly.

Same Poly Bug, 15 salmon later. The uglier, the better (I specially kept this heavily used fly for this chapter)

A few last words

In this chapter I have given you a lot of tips and suggestions but I would like to finish this story with some extra guidelines that will help to increase your catches while you are using or trying dry flies for Atlantic salmon.

1. Believe in what you do because if you have no faith in dry fly fishing for Atlantic salmon, it's not worth trying it at all.
2. Small rivers with water that contains a certain amount of tannic acid are my favourite by far, but that doesn't mean that crystal clear waters don't work for dry fly fishing. There are many famous gin-clear rivers that are excellent for dry fly fishing.
3. Always start without wading first and fishing close to the bank, and then slowly widen the distance.
4. Fish by sight when possible and go for the active fish like head-and-tailers, or fish that break the surface film regularly. Don't concentrate on jumpers, as so many people do.
5. Concentrate at all times and expect a fish to take your fly when you least expect it.
6. When fish stop rising to your fly, change your fly.
7. Salmon have their lively and very lazy moments, so be ready when the fish get active. Some days the lazy moments are more plentiful, but salmon can be very active for several hours too.
8. When you have problems hooking rising fish, just strike more quickly and set the hook on any swirls or broken surface water underneath your fly.
9. There are many people who have caught salmon by skating dry flies, but we see this as luck compared to the catches we have made in totally dead-drift presentations. We present all our dry flies with the fly in front and with a perfect dead drift.
10. Don't make unnecessary dry and backcasts and when you want to dry your flies in the air never do it over the water you are fishing. We only make one cast; if the cast is wrong, just let it go until the drift is done and start a new one.
11. Fish the feeding lanes more intensively.
12. Avoid extremely long casts because it's hard to get extra slack into your line.
13. Try to locate the lies of the fish that you want to catch and cast your fly about three metres upstream of that spot.
14. Check your tippet after every fish you land, and when fishing rocky areas also check the hook point more frequently as well.
15. Watch the area you are fishing and don't wade too far out. Every year I see fishermen walking in the river and choosing their fishing position exactly at the spot where most salmon usually have their favourite lies. Atlantic salmon take dry flies very often in shallow lies and fast water with a very smooth surface.
16. Other hotspots for dry fly are long, deep and colder pools with hardly any current.

When fishing Poly Bugs, tiny colour differences can have a huge impact on your catches